Recent Research and Application of Amorphous Materials

Recent Research and Application of Amorphous Materials

Guest Editors

Pengwei Li
Yanfei Zhang

MDPI

Basel • Beijing • Wuhan • Barcelona • Belgrade • Novi Sad • Cluj • Manchester

Guest Editors

Pengwei Li
College of Material Science & Chemical Engineering
Harbin Engineering University
Harbin
China

Yanfei Zhang
Shandong Academy of Sciences
Qilu University of Technology
Jinan
China

Editorial Office
MDPI AG
Grosspeteranlage 5
4052 Basel, Switzerland

This is a reprint of the Special Issue, published open access by the journal *Inorganics* (ISSN 2304-6740), freely accessible at: https://www.mdpi.com/journal/inorganics/special_issues/N3I9HRJ253.

For citation purposes, cite each article independently as indicated on the article page online and as indicated below:

Lastname, A.A.; Lastname, B.B. Article Title. *Journal Name* **Year**, *Volume Number*, Page Range.

ISBN 978-3-7258-6822-3 (Hbk)
ISBN 978-3-7258-6823-0 (PDF)
https://doi.org/10.3390/books978-3-7258-6823-0

Contents

About the Editors

Pengwei Li

Pengwei Li holds a PhD in Engineering and currently serves as a Lecturer at Harbin Engineering University and a Master's Supervisor. He received his PhD in Materials Science and Engineering from Northeastern University in October 2024. His research interests focus on new energy materials, with particular emphasis on the recycling and regeneration of spent lithium-ion batteries. To date, he has published more than 30 SCI-indexed papers as the first or corresponding author in leading international journals, including *Chemical Society Reviews, Joule, Matter, Advanced Functional Materials, Energy Storage Materials, Renewable and Sustainable Energy Reviews, Journal of Energy Chemistry, Journal of Hazardous Materials, Green Chemistry*, and *ACS Energy Letters*. His publications have accumulated a total impact factor exceeding 350, an h-index of 25, and more than 2000 citations. Among these works, three papers have been recognized as ESI Highly Cited Papers, and two have been selected as ESI Hot Papers. He currently serves as a junior editor for *Rare Metals, Advanced Powder Materials, Chinese Chemical Letters*, and *Carbon Neutralization*, and is also a member of the editorial board of *Current Analytical Chemistry*.

Yanfei Zhang

Yanfei Zhang is a Professor at the Qilu University of Technology, Jinan, China. She received her Ph.D. degree in Physics from Shandong University, China, in 2009. From 2011 to 2012, she worked as a Postdoctoral fellow at Aalborg University, Denmark. She leads a dynamic research group that focuses on the fundamental and applied aspects of glass science. Her work delves into the complex dynamics of glass relaxation and the thermodynamics of the glass transition. In the field of energy materials, she is dedicated to exploring novel glass electrode materials and high-performance solid electrolytes for all-solid-state lithium-ion batteries.

Preface

The rapid development of amorphous materials has become one of the most dynamic directions in contemporary materials science. Distinguished by the absence of long-range atomic order, amorphous materials exhibit a combination of mechanical, chemical, optical, and functional properties that are difficult to achieve in crystalline systems. These characteristics have driven their increasing importance in fields such as energy storage and conversion, optoelectronics, environmental technologies, and biomedicine.

The motivation for compiling this Reprint arises from the growing need to consolidate recent scientific advances and to provide a coherent perspective on how experimental techniques, theoretical analysis, and computational methods are jointly advancing the understanding and application of inorganic amorphous materials. Significant progress has been made in synthesis strategies, advanced characterization, and modeling approaches, including multiscale simulation and data-driven design, yet the field remains highly interdisciplinary and rapidly evolving.

This Reprint is intended to serve researchers, engineers, and graduate students working in materials science, physics, chemistry, and related engineering disciplines. By bringing together representative studies and critical insights, it aims to support knowledge exchange, stimulate cross-disciplinary collaboration, and provide a reference framework for future research and technological development in amorphous materials.

Pengwei Li and Yanfei Zhang
Guest Editors

inorganics

MDPI

Editorial

Recent Research on the Applications of Amorphous Materials

Pengwei Li

College of Materials Science and Chemical Engineering, Harbin Engineering University, Harbin 150010, China; pengweili2024@hrbeu.edu.cn

Inorganic amorphous materials continue to play a foundational role in modern materials science, enabling advances in photonics, catalysis, electronics, energy storage, and biomedical technologies. Their unique properties arise from the absence of long-range order, the presence of flexible bonding environments, and the ability to tailor defect populations, which together enable mechanical resilience, optical tunability, chemical stability, and multifunctionality. Recent progress in their synthesis and in situ characterization has deepened our understanding of amorphous structures and property relationships, as well as facilitating breakthroughs in metallic glasses, oxide glasses, amorphous semiconductors, and functional thin films [1–3].

This Special Issue of *Inorganics* highlights representative advances across amorphous and inorganic material systems, illustrating how compositional tuning, structural engineering, and defect modulation synergistically enable performance improvements in electronics, energy devices, dental materials, and radiation-resistant systems.

Metallic glasses (MGs) remain a central topic in amorphous materials research due to their high strength, extended elastic limit, and corrosion resistance. Over the past decade, deep insights have been gained into their glass transition behavior, deformation mechanisms, and structural relaxation dynamics [1,2]. Feng et al. provide a comprehensive review of amorphous alloy development, covering formation mechanisms, atomic packing, and emerging applications from aerospace components to catalysis [4]. Their assessment resonates with broader theoretical frameworks on metallic glass processing and glass-forming ability [3,5].

Amigo et al. investigate the dynamic plasticity of CuZr MG/Cu composites under high-velocity impact, showing that crystalline inclusions facilitate shear-band multiplication and energy dissipation [6]. These findings complement atomistic analyses of mechanical instability and shear-transformation-zone (STZ) activation in amorphous alloys [7]. Additionally, Jiang et al. demonstrate that minor Ag additions greatly enhance the corrosion resistance and antibacterial properties of Fe-based bulk metallic glasses (BMGs) while maintaining structural integrity and biocompatibility [8], consistent with studies on micro-patterning and surface modification of BMGs for biomedical applications [9].

Functional inorganic glasses remain indispensable in photonics, ionics, and optoelectronics. Vijayalakshmi et al. reveal strong NIR emission and tunable dielectric behavior in Ni^{2+}-activated $PbO–GeO_2$ glasses, demonstrating their potential for solid electrolytes and nonlinear photonic components [10]. These findings align with recent efforts to develop multi-component glass systems with controlled phase separation, enhanced stability, and tailored optical transitions [3,11–14].

Deng et al. highlight an amorphous phase separation (APS) engineering strategy that suppresses crystallization in $SiO_2–Al_2O_3–P_2O_5–Li_2O–ZrO_2$ glass systems [15]. Their approach echoes broader interest in leveraging nanoscale heterogeneity to enhance glass stability and mechanical strength [12,13].

In amorphous semiconductor systems, Ramos-Serrano et al. analyze SiO_xC_y:H films produced via HW-CVD, demonstrating how oxygen-deficient centers and band-tail states govern photoluminescence behavior [16]. These results complement major developments in amorphous oxide semiconductors (AOS), such as IGZO and related high-mobility amorphous oxides used in next-generation thin-film transistors [17,18], as well as chalcogenide-based amorphous photonic materials for broadband optical modulation [19].

Velandia et al. report high-responsivity microbolometers based on nitrogen-doped amorphous SiGe:H,N frameworks [20], contributing to the growing interest in amorphous semiconductors for low-cost IR sensing and flexible electronics [21].

Understanding amorphization under extreme environments is vital for nuclear materials, space technology, and semiconductor reliability. Xiao et al. demonstrate that helium implantation induces nanocrystalline-to-amorphous transitions in ZrN thin films, with He atoms segregating along grain boundaries to initiate structural collapse [22]. Their atomic-level insights complement theoretical analyses of irradiation-induced amorphization, defect accumulation, and bubble evolution in ceramic and metallic systems [23–25].

These studies underscore the broader importance of interface-dominated amorphization, grain-boundary instability, and defect-driven phase transitions in nanoscale materials, as highlighted in recent reviews on radiation effects in oxides and carbides [23,24].

Dental and biomedical inorganic materials represent another area of rapid progress. Chen et al. review calcium-based remineralization systems (e.g., hydroxyapatite, α/β-TCP, and CPP-ACP), highlighting their antibacterial action and regenerative potential [26]. Their formulations align with recent developments in bioactive glass systems that promote ion release, pH regulation, and enamel regeneration [19,27].

De et al. report fluoride-modified nanotube-reinforced glass ionomer liners with improved antibacterial properties against S. mutans [28]. These findings echo parallel advances in inorganic antimicrobial coatings and hybrid calcium-phosphate materials for dental preservation [29,30].

Further, the work on Ag-modified Fe-based BMGs by Jiang et al. [8] complements the broader research trend of designing amorphous metallic biomaterials with controlled ion release, enhanced corrosion resistance, and multifunctional antibacterial behavior [30,31].

Amorphous materials have emerged as promising candidates for electrochemical energy storage and catalysis due to their abundant active sites and flexible coordination environments. Amorphous oxides and hydroxides have shown excellent performance in supercapacitors, electrocatalytic water splitting, and lithium storage [31–33]. These works provide a complementary backdrop to the functional amorphous materials featured in this Special Issue, expanding the design space for sustainable energy devices.

Together, the contributions in this Special Issue reflect the breadth, depth, and growing sophistication of amorphous materials research. From metallic glasses and radiation-resistant ceramics to photonic glasses and dental materials, the field is being reshaped by multi-scale modeling, high-throughput synthesis, machine-learning-assisted design, and advanced in situ characterization [34]. Future progress will depend on understanding atomic-scale features—short-range order, bonding configurations, defect motifs, and integrating them with macroscopic performance requirements to develop robust, sustainable, high-performance amorphous systems across structural, photonic, biomedical, and energy applications.

As Guest Editors, we sincerely thank all authors and reviewers whose work made this Special Issue possible. We also extend our appreciation to the editorial team of *Inor-*

ganics for their continued support in promoting advances in inorganic and amorphous materials science.

Conflicts of Interest: The authors declare no conflicts of interest.

References

1. Sun, Y.; Concustell, A.; Greer, A. Thermomechanical Processing of Metallic Glasses: Extending the Range of the Glassy State. *Nat. Rev. Mater.* **2016**, *1*, 16039. [CrossRef]
2. Yu, H.; Wang, W.; Samwer, K. The β relaxation in metallic glasses: An overview. *Prog. Mater. Sci.* **2013**, *16*, 183–191. [CrossRef]
3. Schroers, J. Processing of Bulk Metallic Glass. *Adv. Mater.* **2010**, *22*, 1566–1597. [CrossRef]
4. Feng, Z.; Geng, H.; Zhuang, Y.; Li, P. Progress, Applications, and Challenges of Amorphous Alloys: A Critical Review. *Inorganics* **2024**, *12*, 232. [CrossRef]
5. Inoue, A.; Takeuchi, A. Recent Progress in Bulk Glassy Alloys. *Mater. Trans.* **2002**, *43*, 1892–1906. [CrossRef]
6. Amigo, N.; Wachter, J.; Leiva, P. Impact-Induced Plastic Deformation in CuZr Metallic Glass and MG/Cu Composites. *Inorganics* **2025**, *13*, 141. [CrossRef]
7. Cheng, Y.; Ma, E. Atomic-Level Structure and Structure–Property Relationship in Metallic Glasses. *Prog. Mater. Sci.* **2011**, *56*, 379–473. [CrossRef]
8. Jiang, L.; Fan, X.; Li, Q.; Li, X.; Jiang, T.; Wei, Q. Enhanced Antimicrobial and Biomedical Properties of Fe-Based Bulk Metallic Glasses Through Ag Addition. *Inorganics* **2025**, *13*, 105. [CrossRef]
9. Liu, L.; Hasan, M.; Kumar, G. Metallic Glass Nanostructures: Fabrication, Properties, and Applications. *Nanoscale* **2014**, *6*, 2027–2036. [CrossRef]
10. Vijayalakshmi, L.; Meera, S.; Vijay, R.; Palle, K.; Ramesh, P.; Kwon, S.; Raju, G. Nickel Ions Activated PbO–GeO_2 Glasses for the Application of Electrolytes and Photonic Devices. *Inorganics* **2024**, *12*, 215. [CrossRef]
11. Venugopal, R.; Bishnoi, S.; Singh, S.; Zaki, M.; Grover, H.; Bauchy, M.; Agarwal, M.; Krishnan, N. Artificial Intelligence and Machine Learning in Glass Science and Technology: 21 Challenges for the 21st Century. *Int. J. Appl. Glass Sci.* **2021**, *121*, 277–292. [CrossRef]
12. Wisitsorasak, A.; Wolynes, P. On the Strength of Glasses. *Proc. Natl. Acad. Sci. USA* **2012**, *109*, 16068–16072. [CrossRef]
13. Tipeev, A.; Zanotto, E.; Rino, J. Crystal Nucleation Kinetics in Supercooled Germanium: MD Simulations Versus Experimental Data. *J. Phys. Chem. B* **2020**, *124*, 7979–7988. [CrossRef]
14. Hirotsu, Y.; Hanada, T.; Ohkubo, T.; Makino, A.; Yoshizaw, Y.; Nieh, T. Nanoscale Phase Separation in Metallic Glasses Studied by Advanced Electron Microscopy Techniques. *Intermetallics* **2004**, *12*, 1081–1088. [CrossRef]
15. Deng, M.; Wang, M.; Rao, Y.; Xu, Y.; Wu, D.; Lin, S.; Lu, P. Discovering Novel Glass with Robust Crystallization Resistance via Amorphous Phase Separation Engineering. *Inorganics* **2024**, *12*, 149. [CrossRef]
16. Ramos, J.; Matsumoto, Y.; Ávila, A.; Romero, G.; Meneses, M.; Morales, A.; Luna, J.A.; Flores, J.; Minquiz, G.; Moreno, M. Luminescence Study of Hydrogenated Silicon Oxycarbide (SiO_xC_y:H) Thin Films Deposited by Hot Wire Chemical Vapor Deposition as Active Layers in Light Emitting Devices. *Inorganics* **2024**, *12*, 298. [CrossRef]
17. Robertson, J. High Dielectric Constant Oxides. *Eur. Phys. J.-Appl. Phys.* **2004**, *28*, 265–291. [CrossRef]
18. Medvedeva, J.; Buchholz, D.; Chang, R. Recent Dvances in Understanding the Structure and Properties of Amorphous Oxide Semiconductors. *Adv. Electron. Mater.* **2017**, *3*, 1700082. [CrossRef]
19. Savage, J. Optical Properties of Chalcogenide Glasses. *J. Non-Cryst. Solids* **1982**, *47*, 101–115. [CrossRef]
20. Velandia, O.; Torres, A.; Morales, A.; Hernández, L.; Luna, A.; Monfil, K.; Flores, J.; Minquiz, G.; Jiménez, R.; Moreno, M. Uncooled Microbolometers Based on Nitrogen-Doped Hydrogenated Amorphous Silicon-Germanium (a-SiGe:H,N). *Inorganics* **2025**, *13*, 126. [CrossRef]
21. Robertson, J. Deposition Mechanism of Hydrogenated Amorphous Silicon. *J. Appl. Phys.* **2000**, *87*, 2608–2617. [CrossRef]
22. Xiao, X.; Sun, S.; Jiang, W.; Qin, X.; Liu, Q.; Lao, Y. Nanocrystalline–Amorphous Transition in ZrN Nanofilms Induced by Helium Accumulation at Grain Boundaries. *Inorganics* **2025**, *13*, 158. [CrossRef]
23. Banhart, F. Structural Transformations in Carbon Nanoparticles Induced by Electron Irradiation. *Phys. Solid State* **2002**, *44*, 399–404. [CrossRef]
24. Matsunaga, A.; Kinoshita, C.; Nakai, K.; Tomokiyo, Y. Radiation-Induced Amorphization and Swelling in Ceramics. *J. Nucl. Mater.* **1991**, *179–181*, 457–460. [CrossRef]
25. Khiara, N.; Coulombier, M.; Raskin, J.; Bréchet, Y.; Pardoen, T.; Onimus, F. Helium Nano-Bubbles in Copper Thin Films Slows Down Creep Under Ion Irradiation. *Acta Mater.* **2025**, *288*, 120854. [CrossRef]

26. Chen, J.; Zhang, Y.; Yin, I.X.; Yu, O.Y.; Chan, A.K.Y.; Chu, C.H. Preventing Dental Caries with Calcium-Based Materials: A Concise Review. *Inorganics* **2024**, *12*, 253. [CrossRef]
27. Jones, J. Review of Bioactive Glass: From Hench to Hybrids. *Acta Biomater.* **2013**, *9*, 4457–4486. [CrossRef]
28. De, J.; Valdés, S.; Morales, A.; Jiménez, L.; Correa, R.; Lara, E.; Toral, V.; Chanes, O.; Velázquez, U.; Morales, R. The Growth-Inhibitory Effect of Glass Ionomer Liners Reinforced with Fluoride-Modified Nanotubes. *Inorganics* **2025**, *13*, 190. [CrossRef]
29. Vasluianu, R.; Bobu, L.; Lupu, I.; Antohe, M.; Bulancea, B.; Moldovanu, A.; Stamatin, O.; Holban, C.; Dima, A. Innovative Smart Materials in Restorative Dentistry. *J. Funct. Biomater.* **2025**, *16*, 318. [CrossRef]
30. Wang, C.; Makvandi, P.; Zare, E.; Tay, F.; Niu, L. Advances in Antimicrobial Organic and Inorganic Nanocompounds in Biomedicine. *Adv. Ther.* **2020**, *3*, 2000024. [CrossRef]
31. Yan, S.; Abhilash, K.; Tang, L.; Yang, M.; Ma, Y.; Xia, Q.; Guo, Q.; Xia, H. Research Advances of Amorphous Metal Oxides in Electrochemical Energy Storage and Conversion. *Small* **2019**, *15*, 1804371. [CrossRef] [PubMed]
32. Guo, C.; Shi, Y.; Lu, S.; Yu, Y.; Zhang, B. Amorphous nanomaterials in electrocatalytic water splitting. *Chin. J. Catal.* **2021**, *42*, 1287–1296. [CrossRef]
33. Ji, S.; Kim, S.; Song, W.; Yoon, Y.; Myung, S.; Lim, J.; Jung, H.; Choi, S.; An, K.; Lee, S. Extraordinary Lithium Storage Capacity and Lithiation Mechanism of Partially Amorphous Molybdenum Sulfide on Chemically Exfoliated Graphene. *Electrochim. Acta* **2020**, *354*, 136636. [CrossRef]
34. Falqui, A.; Loche, D.; Casu, A. In Situ TEM Crystallization of Amorphous Iron Particles. *Crystals* **2020**, *10*, 41. [CrossRef]

inorganics

MDPI

Article

The Growth-Inhibitory Effect of Glass Ionomer Liners Reinforced with Fluoride-Modified Nanotubes

Ricardo De Jesús-Pascual [1], Elias Nahum Salmerón-Valdés [1,*], Adriana Alejandra Morales-Valenzuela [1], Leticia Verónica Jiménez-Rojas [2], Rodrigo Correa-Prado [3], Edith Lara-Carrillo [1], Víctor Hugo Toral-Rizo [1], Osmar Alejandro Chanes-Cuevas [4], Ulises Velázquez-Enríquez [1], Raúl Alberto Morales-Luckie [5] and Javier Jaramillo-García [1]

1 Center for Research and Advanced Studies in Dentistry, School of Dentistry, Autonomous University of Mexico State, Toluca 50000, Estado de Mexico, Mexico
2 Infectious Diseases Research Unit, Hospital Infantil de Mexico Federico Gómez, Mexico City 06720, Mexico
3 Tissue Bioengineering Laboratory, Postgraduate Studies and Research Division, Faculty of Dentistry, National Autonomous University of Mexico, Mexico City 04510, Mexico
4 Dental Biomaterials Laboratory, Postgraduate Division, Dental School, National Autonomous University of Mexico, Mexico City 04510, Mexico
5 Joint Research Center for Sustainable Chemistry, UAEM-UNAM, Toluca 50200, Estado de Mexico, Mexico
* Correspondence: salmeron81@hotmail.es

Abstract: The aim of this research was to compare the growth-inhibitory effect of halloysite-based nanotubes preloaded with sodium fluoride incorporated into two commercial glass ionomers (Vitrebond 3M^{TM} and Ionobond VOCO) for indirect pulp capping. Methods: Sixty samples were prepared and were distributed into two control groups, two positive control groups and two experimental groups. A total of 10% of the total weight of ionomer powder required to prepare each sample was replaced with nanotubes that had been preloaded at 2000 parts per million (minimum inhibitory dose for Streptococcus mutans established in this study using the McFarland index). The growth-inhibitory effect was determined by placing the samples in Petri dishes inoculated with *S. mutans* for 24 h at 37 °C. Results: Regarding the control groups, only Vitrebond demonstrated a growth inhibition zone; both experimental groups showed an inhibitory effect, and statistical differences were observed when the experimental Ionobond group and control groups were compared. Conclusions: The ionomers reinforced with fluorine-modified nanotubes showed an adequate inhibitory effect on *Streptococcus mutans.*

Keywords: glass ionomer cements; fluoride; halloysite; nanotubes; *Streptococcus mutans*; antibacterial effect

1. Introduction

The principal challenge of modern restorative dentistry is the protection and preservation of pulpal vitality with the objective to avoid a root canal treatment; in deep caries lesions, the management includes the removal of damaged tissue and the subsequent application of a liner of certain biomaterials. Liners are a relatively thin indirect pulp capping (IPC) that provides a barrier to protect dentin in deep cavity preparations with proximity to the pulp if no exposure of this tissue is detected [1,2]. They eliminate or reduce postoperative hypersensitivity in resin-based composite restorations [3]. They are widely used to provide an adequate biological seal and cariostatic action [4]. Therefore, the principal objective of any indirect pulp capping procedure is to avoid bacterial filtration to guarantee the health, vitality and protection of the pulp. Conventional glass ionomer cements (GICs) are the most used materials for IPC treatments, and they have been shown to reduce shrinkage stress and the formation of spaces in the dentin/resin adhesive interface [2,5,6]. Glass ionomer-based dental materials exhibit several advantages, the most significant of which include their fluoride-releasing capacity and biocompatibility with the dental pulp. Furthermore, these cements demonstrate low cytotoxicity, the ability to form chemical bonds with dental hard tissues, a coefficient of thermal expansion comparable to that of natural tooth structures, and bacteriostatic properties [7–11].

On the other hand, some disadvantages like difficult handling, low compressive and flexural strengths and low fracture resistance have been reported [12]. In recent years, RMGIC (resin-modified glass ionomer cement) has been developed to eliminate said disadvantages, and it has been used in IPC treatments [3,13–15].

Previous studies reported that both GICs and RMGICs could stimulate the remineralization of dentin affected by caries and contribute to the prevention of secondary caries, which is considered the main cause of failure of resin-based restorations [16,17]. The bacteriostatic function of glass ionomer has been attributed to the fluoride release ability of this material.

Fluoride is an antibacterial material that strengthens the structure of enamel, making it less susceptible to demineralization and caries formation due to the generation of fluorapatite crystals and the promotion of enamel remineralization; these mechanisms result in less soluble tissue in the acidic enamel environment that is conducive to the growth of microorganisms such as *Streptococcus mutans* (*S. mutans*) [18–20].

This microorganism inhabits the human oral cavity, specifically in the biofilm of dental surfaces (dental plaque), and its cariogenic potential is characterized by several key traits: rapid conversion of carbohydrates into lactic acid (acidogenicity); the ability to synthesize large amounts of glucan in the presence of sucrose; a strong affinity for colonizing the hard tissues of the tooth; the capacity to produce an extracellular polymeric matrix; and the ability to survive and proliferate in low pH environments (aciduricity).Although *S. mutans* is not the only microorganism that causes dental caries, it has been shown to be capable of creating a favorable environment for other aciduric and acidogenic species [21,22].

Regarding the fluoride released by GICs and RMGIs, there is controversy about the true antibacterial capacity of these cements because almost all studies have reported a fluoride release level that does not exceed 50 ppm [23–26]. Morales et al. (2020) reported that glass ionomers release approximately 17.8 ppm fluoride [27]. On the other hand, some studies mentioned that the minimum quantity required to inhibit the growth of *S. mutans* is between 4000 and 5000 ppm. Pradiptama et al. (2019) stated that the minimum concentration of sodium fluoride necessary to inhibit the growth of *S. mutans* is 4800 ppm (4.8 mg/mL) [28]. Therefore, it can be said that the antibacterial effect of fluoride-releasing materials is directly related to the quantity and duration of the fluoride release, and for this reason, the necessary quantity of fluoride to inhibit bacterial growth cannot be obtained from GICs or RMGIs.

Several studies have focused on improving the physical, mechanical and antibacterial properties of ionomers by using nanotechnology [29]. Some studies reported an improvement in the microhardness, compressive strength, shear bonding and flexural strength of GICs with the incorporation of nanoparticles into the glass ionomer powder [30–32]. However, there has been great interest in developing or modifying some GICs with nanostructures like nanoparticles (zinc oxide, graphene–silver and magnesium oxide nanoparticles) to enhance their antibacterial capacity [33–35].

Nanotubes are nanostructures with drug delivery ability; drugs can be loaded inside nanotubes, and they can improve the antibacterial properties and increase the time of action of different compounds [36,37]. There are nanotubes made of different materials, such as carbon, trititanate, graphene, boron and halloysite. Halloysite nanotubes (HNTs) are a low-cost, nanometer-sized porous tubular structure with a lumen diameter of 10–15 nm, which allows for the encapsulation of drugs or active elements, making them a nanometric container capable of releasing said drugs in a controlled manner. The outer diameter of nanotubes is approximately of 50–80 nm, and they have a length of 1 μm [38,39]. Recent research has shown that HNTs can be loaded with various antimicrobial agents [36]. Other advantages of halloysite nanotubes are their high mechanical strength, thermal stability and adequate biocompatibility [40,41].

The antibacterial properties of glass ionomer cements have been evaluated in previous studies via the agar diffusion method, which has the advantages of simplicity, low cost, easy interpretation of results and the possibility of analyzing a large number of antimicrobials and microorganisms [42–46]. The process consists of inoculating a microorganism onto an agar plate, and if the antimicrobial agent is effective, it will diffuse into the agar and be able to inhibit the growth of the microorganisms, and the diameter of the zone of growth inhibition around the sample is measured.

Therefore, this study aimed to load HNTs with a minimum inhibitory dose of sodium fluoride and to determine the growth-inhibitory effect of these nanotubes on *S. mutans* when incorporated into a GIC and an RMGIC.

2. Results

Fluoride and halloysite nanotubes were analyzed through Fourier transform infrared spectroscopy. In Figure 1a, vibrations can be observed between 1000 and 1400 cm^{-1}, characteristic of fluorinated compounds (F), and the vibrations observed between 300 and 500 cm^{-1} correspond to the interaction between sodium and fluorine ions. Regarding the halloysite nanotubes (b), the flexural vibrations between 400 and 600 cm^{-1} correspond to the presence of silicate groups (Si-O). The vibrations between 1000 and 1200, characteristic of silicate groups, and the flexural vibrations between 600 and 700 cm^{-1}, typical of minerals containing aluminum (Al-O) in their structure, confirmed the presence of halloysite.

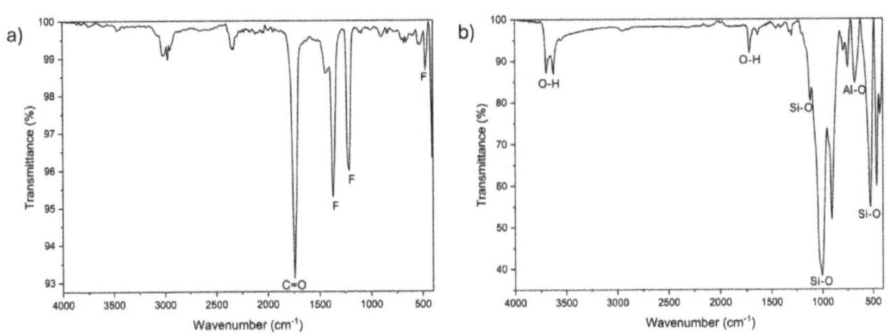

Figure 1. Fourier transform infrared spectroscopy of fluoride (**a**) and halloysite nanotubes (**b**).

As shown in Figure 2a,b, the corresponding infrared spectrum for the control groups showed functional groups belonging to OH groups between the frequencies of 3200–3600 cm^{-1} and the presence of aliphatic chains (C-H) between 2850 and 2960 cm^{-1}. Bands related to silicate groups (Si-O) were observed at frequencies of 1075 for Ionobond and 1052 for Vitrebond, and the bands at frequencies of 732 for Ionobond and 723 for Vitrebond, indicative of aluminum groups (Al-O), correspond to the presence of halloysite in the positive control groups. For the experimental groups, the presence of the groups mentioned previously was observed, and the appearance of fluoride was confirmed via the band peaks at 1178 and 1456 for Ionobond and at 1143 and 1416 for Vitrebond.

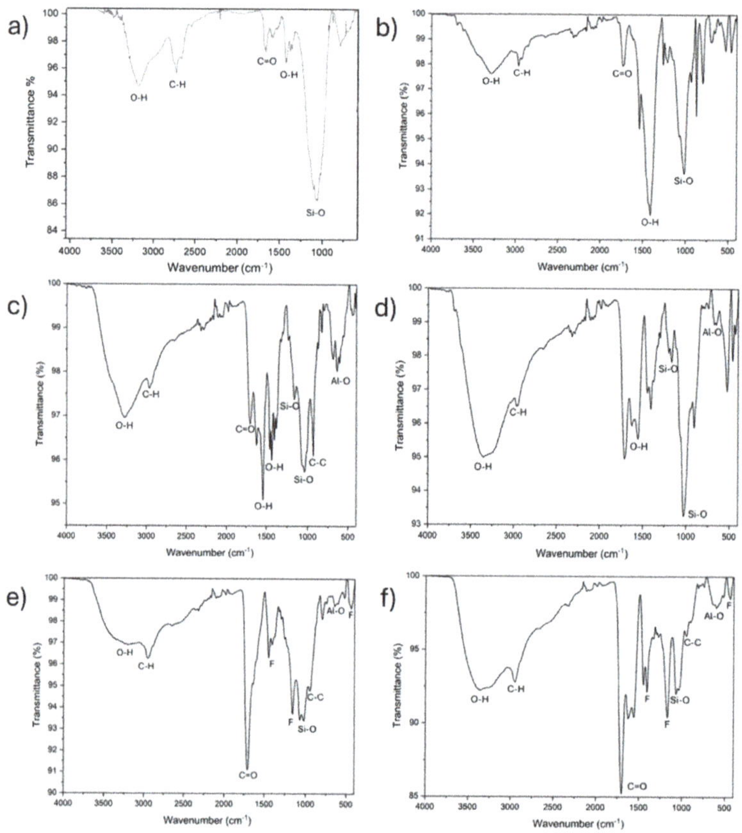

Figure 2. Fourier transform infrared spectroscopy of study groups. The control groups of (**a**) Ionobond (IB) and (**b**) Vitrebond (VB) and the positive control groups. (**c**) Ionobond with nanotubes (IBHNT), (**d**) Vitrebond with nanotubes (VBHNT), (**e**) Ionobond with fluoride-modified nanotubes (IBHNT-NaF) and (**f**) Vitrebond with fluoride-modified nanotubes (VBHNT-NaF).

Regarding the growth-inhibitory effect, the minimum and maximum inhibitory concentrations of the control group were 3000 and 4000 ppm, respectively, while for the experimental group, the concentrations were 2000 and 4000 ppm (Figure 3).

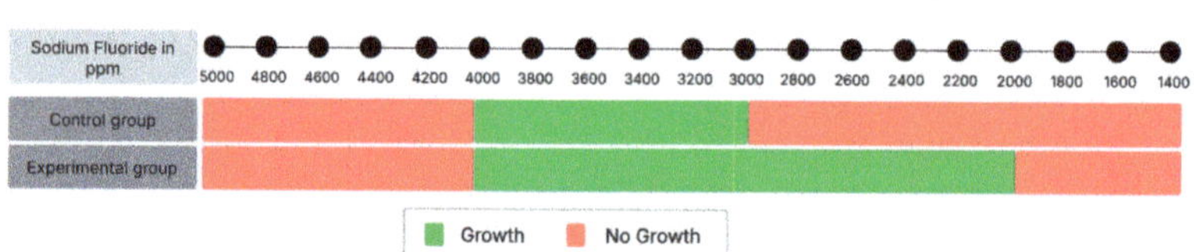

Figure 3. Inhibitory concentrations of control (NaF) and experimental (HNTs-NaF) groups, shown in ppm (parts per million).

Among the control and positive control groups, only VB and VBHNT showed a bacterial growth inhibition zone of 8.3 and 8.5 mm, respectively, for *S. mutans* (Figure 4).

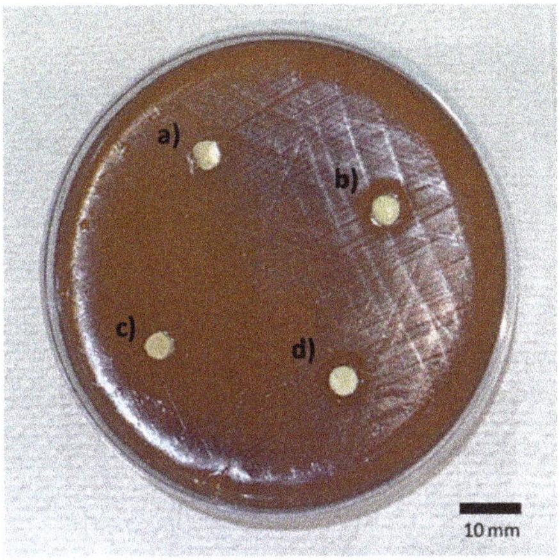

Figure 4. Growth inhibition zone of control and positive control groups. (a) Ionobond (IB) and (b) Vitrebond (VB), which formed the control group; (c) Ionobond with halloysite nanotubes (IBHNT) and (d) Vitrebond with halloysite nanotubes (VBHNT), which formed the positive control group.

In Figure 5, it can be observed that all of the experimental groups presented an antibacterial effect on the microorganism analyzed in this study. An average bacterial growth inhibition zone of 8.60 ± 0.51 mm was observed for the IBHNT-NaF, while the VBHNT-NaF showed a greater inhibition zone than the control, with a mean of 9.10 ± 0.56 mm.

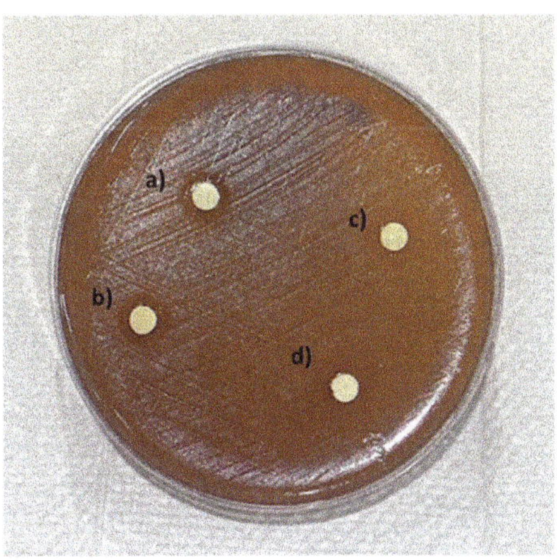

Figure 5. Growth inhibition zone of experimental groups: (a,b) IB with HNT-NaF at 2000 ppm, (c,d) VB with HNT-NaF at 2000 ppm.

As shown in Table 1, significant differences were observed after carrying out a non-parametric Kruskal–Wallis test when the control and experimental groups were compared ($p = 0.001$). This test is carried out to determine if there are statistically relevant differences between two or more groups with a p value ≤ 0.05. Nevertheless, with the Kruskall–Wallis test, it was not possible to determine between which groups statistically significant differences were present. Therefore, a pairwise comparison test (Mann–Whitney U) was performed to determine between which groups these significant differences were found. The control and positive control groups for Ionobond showed no inhibitory effect, while the control and positive control groups for Vitrebond presented a similar inhibitory effect, whereby no statistically significant differences were observed between these groups. Significant differences were observed between the control groups for Ionobond and Vitrebond ($p = 0.010$) because Vitrebond presented an inhibitory effect of approximately 8.30 mm, whereas Ionobond did not present said inhibitory effect. When Ionobond was compared with both experimental groups, a significant difference was observed because the VBHNT-NaF and IBHNT-NaF experimental groups presented an inhibitory effect of 9.10 ($p = 0.0001$) and 8.60 mm ($p = 0.001$), respectively. Nevertheless, between the control group for Vitrebond and the VBHNT-NaF and IBHNT-NaF experimental groups, no differences were observed (Table 2).

Table 1. Kruskal–Wallis' test comparing the antibacterial effect between the experimental and control groups analyzed.

Groups	Mean (SD)
Control group: Vitrebond (VB)	8.30 (0.483)
Experimental group: (VBHNT-NaF)	9.10 (0.568)
Positive control group Vitrebond: (VBHNT)	8.50 (0.527)
Control group: Ionobond (IB)	0.00 (0.00)
Positive control group Ionobond: (IBHNT)	0.00 (0.00)
Experimental group: IBHNT-NaF	8.60 (0.516)
Kruskal–Wallis' test	$p = 0.001$ *

Mean: average of growth inhibition zone in millimeters; SD: standard deviation; *: $p \leq 0.05$.

Table 2. Mann–Whitney U test.

Groups	Statistics	Statistics Deviation	p Value
IB–VB	25.200	3.402	0.010 *
IB–IBHNT	0.000	0.000	1.000
IB–VBHNT	−27.100	−3.658	0.004 *
IB–IBHNT-NaF	−29.000	−3.915	0.001 *
IB–VBHNT-NaF	−38.700	−5.224	0.0001 *
IBHNT–VBHNT	−27.100	−3.658	0.004 *
IBHNT–IBHNT-NaF	−29.000	−3.915	0.001 *
IBHNT–VBHNT-NaF	−38.700	−5.224	0.0001 *
VB–IBHNT	25.200	3.402	0.010 *
VB–VBHNT	−1.900	−0.256	1.000
VB–IBHNT-NaF	−3.800	−0.513	1.000
VB–VBHNT-NaF	−13.500	−1.822	1.000
VBHNT–IBHNT-NaF	−1.900	−0.256	1.000
VBHNT–VBHNT-NaF	−11.600	−1.566	1.000
IBHNT-NaF–VBHNT-NaF	9.700	1.309	1.000

IB, VB: control groups; IBHNT-NaF, VBHNT-NaF: experimental groups; *: $p \leq 0.05$.

3. Discussion

The anticariogenic effect presented by fluoride-releasing materials is directly related to the duration and amount of fluoride released [27,47]. Glass ionomers are among the dental materials that deliver the highest fluoride release. Previous studies have reported fluoride release amounts of between 17.4 and 32.6 ppm [27,28], which correspond to bacterial inhibition [48]. However, one study reported that a value of between 3040 and 5700 ppm of NaF is needed to eliminate bacteria in the oral cavity [49]; another study reported that a value of between 4000 and 6000 ppm of NaF is needed [50]. Another study reported that the minimum inhibitory amount of NaF necessary for a bactericidal effect on *S. mutans* is 4800 ppm [28].

Regarding the maximum inhibitory concentration of fluoride, it has been established by previous studies that excessive concentrations of fluoride can cause a regrowth because *S. mutans* generates a type of adaptation due to a change in the enolase enzyme inhibitor, which is irreversible and subsequently allows *S. mutans* to stay alive [28,49]

Several studies have tried to enhance the antibacterial characteristics of glass ionomers by incorporating different nanomaterials [30,51,52]. However, some of these studies did not report favorable results [36,53–55].

Nanotubes are nanostructures with the ability to encapsulate drugs; specifically, HNTs have been shown to be suitable for drug transport and prolonged release [54,56]. Massaro et al. [57]. concluded that HNTs do not present bacterial inhibition themselves, which was corroborated in this study, since the positive control groups did not show differences in their inhibitory effects from the control groups. However, HNTs are characterized by an excellent carrying capacity and ability to release and transport drugs [54]. These qualities have positively improved some characteristics of dental materials [53].

In this study, the inhibitory effect of NaF was analyzed at concentrations between 1000 and 10,000 ppm; the established maximum inhibitory concentration (4000 ppm) and minimum inhibitory concentration (3000 ppm) of NaF on *S. mutans* was similar to previous studies [49,50]. The positive control groups analyzed in this study did not show differences with respect to the control groups and were included only to guarantee that the inhibitory effect of the experimental groups was due to the presence of fluorine inside the nanotubes. No bacterial growth inhibition zone was observed in the Ionobond control groups analyzed in this study, so an insufficient amount of NaF was released by this glass ionomer to inhibit *S. mutans*. In the VB control group, a bacterial growth inhibition zone was observed. A previous study suggested that this antibacterial effect could be caused by the release of the initiator diphenyleneiodonium chloride, which can inhibit fibroblast growth and could also inhibit the growth of *S. mutans* [47]. Another study showed that compared to GICs, VB is strongly cytotoxic in different cell cultures [58].

In the present investigation, the antibacterial properties were improved in both experimental groups of glass ionomers modified with HNTs preloaded with NaF. In the IB control group, no antibacterial capacity was observed prior to the incorporation of HNTs, but an inhibitory effect was obtained after the incorporation of nanotubes, and the growth inhibition effect of the VBHNT-NAF experimental group increased with respect to VB because a broader bacterial growth inhibition zone was observed (1 mm approximately). However, no statistically significant differences were observed between these groups (Table 2).

With respect to the minimum and maximum concentrations of NaF required to inhibit the growth of *S. mutans*, HNTs promoted the antibacterial activity of NaF and expanded the bacterial inhibition interval of 3000 to 6000 ppm reported by previous studies [28,49]. However, in the experimental groups that were analyzed in our study, a growth inhibition zone for *S. mutans* was observed from 2000 ppm. These results demonstrate the ability of HNTs to promote NaF release and increase the antibacterial effect of glass ionomers, which is consistent with previous studies demonstrating the efficacy of HNTs loaded with other antimicrobial agents [54,56].

The ability of glass ionomer cements to inhibit bacterial growth as a consequence of their fluoride release has been critically evaluated in this study, because previous studies concluded that the inhibitory effect occurs from 5000 ppm [18]. Taking into account the amount of fluoride released by glass ionomer cements reported in previous studies, which is no higher than 50 ppm [17,18,40], these materials will release an insufficient amount of fluoride to guarantee an environment free of microorganisms like *S. mutans*. Considering this limitation, the clinical application of glass ionomer liners reinforced with fluoride-modified nanotubes could increase the success of indirect pulp capping therapy and decrease the probability of secondary caries, guaranteeing an aseptic environment. On the other hand, the amount of fluoride required to eliminate the *S. mutans* present in dental cavities will be reduced to 2000 ppm due to the capability of HNTs to improve the effect of drugs.

4. Materials and Methods

4.1. Determination of the Inhibitory Effect of Sodium Fluoride (NaF)

The minimum and maximum inhibitory concentrations were determined as follows. A total of 2.2106 g of 99.99% extra pure sodium fluoride (Fagalab; Mocorito, México) was weighed with an analytical balance (Shimadzu Scientific Instruments; Kyoto, Japan) and diluted in deionized water (100 mL) to obtain a solution at a concentration of 10,000 ppm [59]. From this stock solution, solutions were prepared with different concentrations with an interval of 1000 ppm (1000–10,000 ppm), and these were used as the control group. A total of 1 mL of each NaF solution was inoculated with 1 mL of *S. mutans* in Mueller–Hinton broth (0.5 McFarland corresponds 1.5×10^8 CFU/mL) at 37 °C for 24 h in 5 mL test tubes, and this procedure was performed in triplicate (30 samples). For the experimental groups, 10 mg of HNTs preloaded with NaF was directly added to 1 mL aliquots of the prepared solutions with *S. mutans* in Mueller–Hinton broth with the same number of samples. Once the minimum inhibitory concentration was established, the HNTs for incorporation into the ionomers were modified with that concentration of NaF (Figure 6).

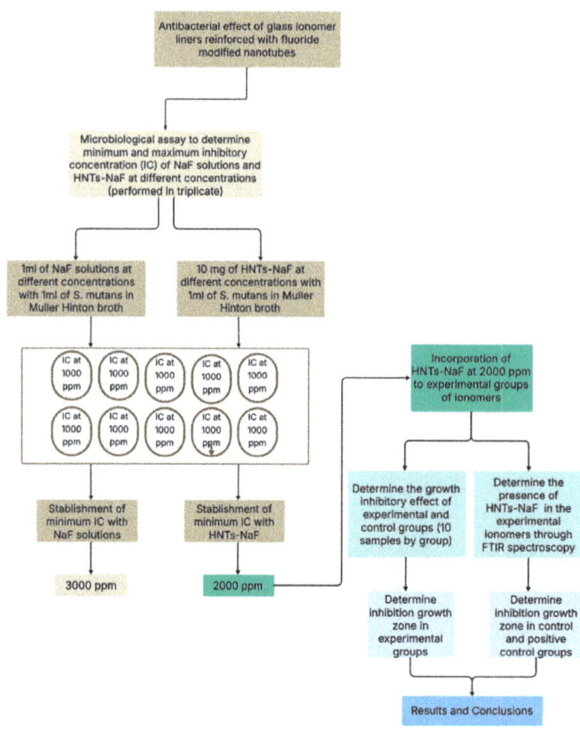

Figure 6. Sketch of experimental procedures.

4.2. Inccorporation of Sodium Fluoride into Halloysite Nanotubes

One gram of halloysite nanotubes (Sigma–Aldrich; St. Louis, MO, USA) was weighed using the previously mentioned analytical balance. The HNTs were immersed in a prepared solution with 98% 3-(trimethoxysilyl) propyl-methacrylate (Sigma–Aldrich) diluted to 5% in 95% acetone (Sigma–Aldrich) for 24 h at a temperature of 110 °C in a drying oven (HERAtherm oven Thermo Fisher Scientific; Waltham, MA, USA).

Subsequently, these nanotubes were incorporated into 10 mL of NaF (Fagalab) at 2000 ppm (the minimum inhibitory concentration established in this study in the above-described experiments) based on the methodology of previous studies, where between 1 and 1.25 g of HNTs was mixed with 5 to 10 mL of antiseptic solution [42,54,60]. Afterwards, 10 mL of pure ethanol at 95% was added, and this solution was sonicated during 1 h. After this procedure, the solution was taken into a drying oven for 10 days at a temperature of 30 °C to remove the residual solvent and obtain the nanotubes loaded with sodium fluoride (HNT-NaF) [61].

4.3. Characterization of the Samples Using FTIR (Fourier Transform Infrared Spectroscopy)

To determine the presence of fluoride in the experimental groups, Fourier transform infrared analysis was used. The glass ionomer cements with fluoride nanotubes were analyzed using a spectrometer (6700 FTIR Perkin Elmer; Waltham, MA, USA) with attenuated total reflectance, employing a diamond/zinc-selenide crystal plate. Each sample was analyzed by thirty-two scans with a 5 cm^{-1} spectral resolution using an infrared spectrum range between 400–4000 cm^{-1}.

4.4. Modification of Glass Ionomers with Preloaded Fluoride Halloysite Nanotubes

For this study, 2 glass ionomers, Vitrebond (VB; 3M ESPE, St. Paul, MN, USA) and Ionobond (IB; VOCO GmbH, Cuxhaven, Germany), were used. Sixty blocks were fabricated in a Teflon matrix with a 3 mm diameter and a thickness of 1 mm. The materials were manipulated according to the instructions provided by the makers. The materials activated by light were polymerized with an LED lamp (Elipar; 3M ESPE) for 20 s; a radiometer (Demetron; Kavo Kerr, Orange, CA, USA) was necessary to confirm the correct intensity of the emitted light. This had to be greater than 400 milliwatts per square centimeter (mW/cm^2).

Using the spoon provided by the manufacturer, the amount of powder indicated was retrieved from the bottle and subsequently weighed. This procedure was performed 10 times to calculate the average amount of powder corresponding to the spoon provided by the manufacturer, this procedure was repeated to elaborate each sample. The average for VB was 0.0483 mg, while the average for IB was 0.1043 mg. Then, 10% of the ionomer powder was replaced with HNTs with and without NaF to prepare the positive control and experimental groups, respectively. The total powder of these groups was mixed with the liquid following the instructions proportionated by the manufacturer.

The control group consisted of 10 blocks of each ionomer analyzed in this study. For the positive control group, 20 blocks of ionomers with HNTs were fabricated and divided into 2 groups (10 blocks per group) as follows: the VBHNT group (VB ionomer with HNTs) and the IBHNT group (IB ionomer with HNTs).

The experimental groups consisted of 10 blocks of each ionomer with HNT-NaF at 2000 ppm, which were distributed into 2 groups (VBHNT-NaF and IBHNT-NaF). The sampling distribution is presented in Figure 7.

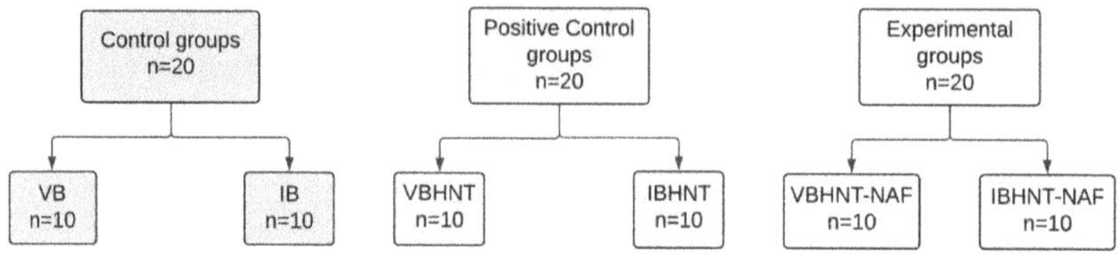

Figure 7. Sample distribution diagram, (n) sample size.

4.5. Microbiological Assay

The microbiological tests were carried out according to the CLSI-established guidelines, specifically those in the document M100 [62]. Mueller–Hinton agar was supplemented with sheep blood at a percentage of 5% (BD Columbia, Heidelberg, Germany), and this was used to seed *S. mutans* (ATCC) 33688 through the cross-streak method, and the samples were subsequently incubated for 24 h at 37 °C.

A bacterial suspension was prepared with a turbidity of 0.5 according to the McFarland scale (equivalent to 1.5×10^8 CFU/mL) in 0.9% $NaCl_2$ solution. The quantification of the bacteria was carried out with the plate count method. The AMH plates were fully inoculated with *S. mutans* without leaving any free zones, and the samples corresponding to the controls and experimental glass ionomer groups were set on the plates. The plates were taken into an incubator (Thermo Fisher Scientific, Waltham, MA, USA) with an anaerobic atmosphere with 5% CO_2 for 24 h at 37 °C. This procedure was performed in triplicate. Subsequently, the bacterial growth on the plates was evaluated to determine the area of bacterial inhibition, which was measured with a Vernier caliper in millimeters.

4.6. Statistical Analysis

The SPSS 25.0 statistical program (IBM, New York, NY, USA) was used to analyze the data obtained. To perform multiple comparisons between the study groups and determine statistical differences between the study groups, the Kruskal–Wallis statistical test was necessary, and the Mann–Whitney U test was performed to determine between which study groups these statistical differences were found. In all the statistical test used in this study, a p value ≤ 0.05 was considered significant.

5. Conclusions

In this study, glass ionomers were modified by replacing 10% of the average amount of powder required with fluoride-modified halloysite nanotubes (HNT-NaF). The glass ionomer Ionobond modified with HNT-NaF obtained an efficient growth-inhibitory effect on *S. mutans*, which this material did not show previously. Vitrebond had a growth-inhibitory effect attributable to the methacrylates in its composition. However, this effect was increased by the incorporation of HNT-NaF, showing wider halos. Similarly, it was established that the minimum amount of sodium fluoride required to inhibit *S. mutans* was between 3000 and 6000 ppm, which could be improved to 2000 ppm by using HNTs.

Halloysite nanotubes are a highly promising additive for improving dental materials. Fluoride-modified halloysite nanotubes can promote the inhibitory effect of glass ionomers on *S. mutans*.

Limitations of the Study

In this study, only one bacterial strain was evaluated during the first 24 h. The evaluation of other bacterial strains with observations beyond this period would have been appropriate to evaluate its long-term effectiveness. It would be desirable to carry out more experiments to evaluate the materials' antibacterial activity, such as the growth kinetics of the biofilm, the viability of bacterial cells, and the morphology of bacterial cells. The authors consider that future studies should be developed to evaluate the behavior and antibacterial capabilities of GICs and RMGIs reinforced with fluoride nanotubes under clinical conditions and analyze their physical and mechanical properties as well as their fluoride release.

Author Contributions: Investigation, R.D.J.-P.; Conceptualization, A.A.M.-V.; Software, R.A.M.-L.; Formal Analysis, U.V.-E.; Data Curation, E.L.-C.; Methodology, L.V.J.-R.; Supervision, O.A.C.-C.; Project Administration, V.H.T.-R.; Visualization, R.C.-P.; Validation, J.J.-G.; Writing—Original Draft, Writing—Review and Editing, E.N.S.-V. All authors have read and agreed to the published version of the manuscript.

Funding: This research received no external funding.

Institutional Review Board Statement: Not applicable.

Informed Consent Statement: Not applicable.

Data Availability Statement: The original contributions presented in this study are included in the article. Further inquiries can be directed to the corresponding author.

Acknowledgments: We thank the Center for Applied Physics and Advanced Technology, U.N.A.M., Campus Juriquilla, Queretaro.

Conflicts of Interest: The authors declare no conflicts of interest.

References

1. Soubhagya, M.; Goud, K.M.; Deepak, B.S.; Thakur, S.; Nandini, T.N.; Arun, J. Comparative in Vitro Evaluation of Internal Adaptation of Resin-Modified Glass Ionomer, Flowable Composite and Bonding Agent Applied as a Liner under Composite Restoration: A Scanning Electron Microscope Study. *J. Int. Oral Health* **2015**, *7*, 27–31. [PubMed]
2. Weiner, R. Liners and Bases in General Dentistry. *Aust. Dent. J.* **2011**, *56*, 11–22. [CrossRef] [PubMed]
3. Strober, B.; Veitz-Keenan, A.; Barna, J.A.; Matthews, A.G.; Vena, D.; Craig, R.G.; Curro, F.A.; Thompson, V.P. Effectiveness of a Resin-Modified Glass Ionomer Liner in Reducing Hypersensitivity in Posterior Restorations: A Study from the Practitioners Engaged in Applied Research and Learning Network. *J. Am. Dent. Assoc.* **2013**, *144*, 886–897. [CrossRef] [PubMed]
4. Mitra, S.B. Adhesion to Dentin and Physical Properties of a Light-Cured Glass-Ionomer Liner/Base. *J. Dent. Res.* **1991**, *70*, 72–74. [CrossRef]
5. Azevedo, L.M.; Casas-Apayco, L.C.; Villavicencio Espinoza, C.A.; Wang, L.; Navarro, M.F.D.L.; Atta, M.T. Effect of Resin-Modified Glass-Ionomer Cement Lining and Composite Layering Technique on the Adhesive Interface of Lateral Wall. *J. Appl. Oral Sci.* **2015**, *23*, 315–320. [CrossRef]
6. Rodrigues, M.d.P.; da Cunha, L.S.; Vilela, A.B.F.; Schettini, A.C.T.; de Bragança, G.F.; França, R.; Soares, C.J. Selective Carious Tissue Removal and Glass Ionomer Liner Reduction of Pulp Stress in Bulk Fill Resin Composite Restorations. *Braz. Oral Res.* **2021**, *35*, e119. [CrossRef]
7. Kasraei, S.; Azarsina, M.; Majidi, S. In Vitro Comparison of Microleakage of Posterior Resin Composites with and without Liner Using Two-Step Etch-and-Rinse and Self-Etch Dentin Adhesive Systems. *Oper. Dent.* **2011**, *36*, 213–221. [CrossRef]
8. Güray Efes, B.; Yaman, B.C.; Gümüştaş, B.; Tiryaki, M. The Effects of Glass Ionomer and Flowable Composite Liners on the Fracture Resistance of Open-Sandwich Class II Restorations. *Dent. Mater. J.* **2013**, *32*, 877–882. [CrossRef]
9. Caneppele, T.; Bresciani, E.; da Silva Ávila, D.; Barcellos, D.; Pucci, C. Effect of Lining Materials on Shear Bond Strength for Composite Restorations In Vitro. *Int. J. Periodontics Restor. Dent.* **2017**, *37*, 137–143. [CrossRef]
10. Ranjbar Omrani, L.; Moradi, Z.; Abbasi, M.; Kharazifard, M.J.; Tabatabaei, S.N. Evaluation of Compressive Strength of Several Pulp Capping Materials. *J. Dent.* **2021**, *22*, 41–47. [CrossRef]
11. Alsunbul, H.; Khan, A.A.; Alqahtani, Y.M.; Hassan, S.A.b.; Asiri, W.; Saadaldin, S.; Alharthi, R.; Aldegheishem, A. Using Functionalized Micron-Sized Glass Fibres for the Synergistic Effect of Glass Ionomer on Luting Material. *J. Funct. Biomater.* **2023**, *14*, 550. [CrossRef]
12. Alsunbul, H.; Khan, A.A.; De Vera, M.A.T.; Bautista, L.S.J.; Javed, R. Utilizing an Oxidized Biopolymer to Enhance the Bonding of Glass Ionomer Luting Cement Particles for Improved Physical and Mechanical Properties. *Biomimetics* **2023**, *8*, 347. [CrossRef]

13. De Magalhães, C.S.; Hara, A.T.; Turssi, C.P.; Serra, M.C.; Giannini, M. Microhardness Evaluation around Composite Restorations Using Fluoride-Containing Adhesive Systems. *J. Appl. Oral Sci.* **2005**, *13*, 259–264. [CrossRef]
14. Duque, C.; Negrini, T.D.C.; Sacono, N.T.; Spolidorio, D.M.P.; De Souza Costa, C.A.; Hebling, J. Clinical and Microbiological Performance of Resin-Modified Glass-Ionomer Liners after Incomplete Dentine Caries Removal. *Clin. Oral Investig.* **2009**, *13*, 465–471. [CrossRef]
15. Deepa, V.L.; Dhamaraju, B.; Bollu, I.P.; Balaji, T.S. Shear Bond Strength Evaluation of Resin Composite Bonded to Three Different Liners: TheraCal LC, Biodentine, and Resin-Modified Glass Ionomer Cement Using Universal Adhesive: An in Vitro Study. *J. Conserv. Dent.* **2016**, *19*, 166–170. [CrossRef]
16. Askar, H.; Krois, J.; Göstemeyer, G.; Schwendicke, F. Secondary Caries Risk of Different Adhesive Strategies and Restorative Materials in Permanent Teeth: Systematic Review and Network Meta-Analysis. *J. Dent.* **2021**, *104*, 103541. [CrossRef]
17. Torres, C.R.G.; Mailart, M.C.; Rocha, R.S.; Sellan, P.L.B.; Contreras, S.C.M.; Di Nicoló, R.; Borges, A.B. The Influence of a Liner on Deep Bulk-Fill Restorations: Randomized Clinical Trial. *J. Dent.* **2020**, *102*, 103454. [CrossRef]
18. Yon, M.J.Y.; Gao, S.S.; Chen, K.J.; Duangthip, D.; Lo, E.C.M.; Chu, C.H. Medical Model in Caries Management. *Dent. J.* **2019**, *7*, 37. [CrossRef]
19. ten Cate, J.M. Current Concepts on the Theories of the Mechanism of Action of Fluoride. *Acta Odontol. Scand.* **1999**, *57*, 325–329. [CrossRef]
20. Naik, S.V.; Attiguppe, P.; Malik, N.; Ballal, S. CPP–ACP and Fluoride: A Synergism to Combat Caries. *Int. J. Clin. Pediatr. Dent.* **2019**, *12*, 120–125. [CrossRef]
21. Lemos, J.A.; Palmer, S.R.; Zeng, L.; Wen, Z.T.; Kajfasz, J.K.; Freires, I.A.; Abranches, J.; Brady, L.J. The Biology of Streptococcus Mutans. *Microbiol. Spectr.* **2019**, *7*, 7. [CrossRef]
22. Kovacs, C.J.; Faustoferri, R.C.; Bischer, A.P.; Quivey, R.G. Streptococcus Mutans Requires Mature Rhamnose-Glucose Polysaccharides for Proper Pathophysiology, Morphogenesis and Cellular Division. *Mol. Microbiol.* **2019**, *112*, 944–959. [CrossRef]
23. Kucukyilmaz, E.; Savas, S.; Kavrik, F.; Yasa, B.; Botsali, M. Fluoride Release/Recharging Ability and Bond Strength of Glass Ionomer Cements to Sound and Caries-Affected Dentin. *Niger. J. Clin. Pract.* **2017**, *20*, 226–234. [CrossRef]
24. Hironaka, N.G.L.; Trizzi, J.Q.; Yoshida, N.M.; Cury, J.A.; Tabchoury, C.P.M.; Botelho, J.N.; Pini, N.I.P.; Pascotto, R.C. Use of Sonic Waves in Bubble Formation, Microhardness and Fluoride Release of a High-Viscosity Glass-Ionomer Cement. *J. Investig. Clin. Dent.* **2019**, *10*, e12456. [CrossRef]
25. Bamoussa, A.; Assery, M.; Pani, S. Fluoride Release and Recharge Abilities of Zinc-Reinforced Glass Ionomer Cement in Comparison to Traditional High Strength Glass Ionomers. *Saudi J. Oral Sci.* **2015**, *2*, 69. [CrossRef]
26. Cibim, D.D.; Saito, M.T.; Giovani, P.A.; Borges, A.F.S.; Pecorari, V.G.A.; Gomes, O.P.; Lisboa-Filho, P.N.; Nociti-Junior, F.H.; Puppin-Rontani, R.M.; Kantovitz, K.R. Novel Nanotechnology of TiO_2 Improves Physical-Chemical and Biological Properties of Glass Ionomer Cement. *Int. J. Biomater.* **2017**, *2017*, 7123919. [CrossRef]
27. Morales-Valenzuela, A.A.; Scougall-Vilchis, R.J.; Lara-Carrillo, E.; Garcia-Contreras, R.; Salmeron-Valdes, E.N.; Aguillón-Sol, L. Comparison of Fluoride Release in Conventional Glass-Ionomer Cements with a New Mechanical Mixing Cement. *Oral Health Prev. Dent.* **2020**, *18*, 319–324. [CrossRef]
28. Pradiptama, Y.; Purwanta, M.; Notopuro, H. Antibacterial Effects of Fluoride in Streptococcus Mutans Growth in Vitro. *Biomol. Health Sci. J.* **2019**, *2*, 1. [CrossRef]
29. Najeeb, S.; Khurshid, Z.; Zafar, M.; Khan, A.; Zohaib, S.; Martí, J.; Sauro, S.; Matinlinna, J.; Rehman, I. Modifications in Glass Ionomer Cements: Nano-Sized Fillers and Bioactive Nanoceramics. *Int. J. Mol. Sci.* **2016**, *17*, 1134. [CrossRef]
30. Ching, H.S.; Luddin, N.; Kannan, T.P.; Ab Rahman, I.; Abdul Ghani, N.R.N. Modification of Glass Ionomer Cements on Their Physical-Mechanical and Antimicrobial Properties. *J. Esthet. Restor. Dent.* **2018**, *30*, 557–571. [CrossRef] [PubMed]
31. Kheur, M.; Kantharia, N.; Iakha, T.; Kheur, S.; Husain, N.A.H.; Özcan, M. Evaluation of Mechanical and Adhesion Properties of Glass Ionomer Cement Incorporating Nano-Sized Hydroxyapatite Particles. *Odontology* **2020**, *108*, 66–73. [CrossRef] [PubMed]
32. Noori, A.J.; Kareem, F.A. Setting Time, Mechanical and Adhesive Properties of Magnesium Oxide Nanoparticles Modified Glass-Ionomer Cement. *J. Mater. Res. Technol.* **2020**, *9*, 1809–1818. [CrossRef]
33. Malekhoseini, Z.; Rezvani, M.B.; Niakan, M.; Atai, M.; Bassir, M.M.; Alizade, H.S.; Siabani, S. Effect of Zinc Oxide Nanoparticles on Physical and Antimicrobial Properties of Resin-Modified Glass Ionomer Cement. *Dent. Res. J.* **2021**, *18*, 73. [CrossRef]
34. Noori, A.J.; Kareem, F.A. The Effect of Magnesium Oxide Nanoparticles on the Antibacterial and Antibiofilm Properties of Glass-Ionomer Cement. *Heliyon* **2019**, *5*, e02568. [CrossRef]
35. Chen, J.; Zhao, Q.; Peng, J.; Yang, X.; Yu, D.; Zhao, W. Antibacterial and Mechanical Properties of Reduced Graphene-Silver Nanoparticle Nanocomposite Modified Glass Ionomer Cements. *J. Dent.* **2020**, *96*, 103332. [CrossRef]

36. Barman, M.; Mahmood, S.; Augustine, R.; Hasan, A.; Thomas, S.; Ghosal, K. Natural Halloysite Nanotubes/Chitosan Based Bio-Nanocomposite for Delivering Norfloxacin, an Anti-Microbial Agent in Sustained Release Manner. *Int. J. Biol. Macromol.* **2020**, *162*, 1849–1861. [CrossRef]
37. Ostadhossein, F.; Moitra, P.; Altun, E.; Dutta, D.; Sar, D.; Tripathi, I.; Hsiao, S.-H.; Kravchuk, V.; Nie, S.; Pan, D. Function-Adaptive Clustered Nanoparticles Reverse Streptococcus Mutans Dental Biofilm and Maintain Microbiota Balance. *Commun. Biol.* **2021**, *4*, 846. [CrossRef]
38. Fakhruddin, K.; Hassan, R.; Khan, M.U.A.; Allisha, S.N.; Razak, S.I.A.; Zreaqat, M.H.; Latip, H.F.M.; Jamaludin, M.N.; Hassan, A. Halloysite Nanotubes and Halloysite-Based Composites for Biomedical Applications. *Arab. J. Chem.* **2021**, *14*, 103294. [CrossRef]
39. Yuan, P.; Tan, D.; Annabi-Bergaya, F. Properties and Applications of Halloysite Nanotubes: Recent Research Advances and Future Prospects. *Appl. Clay Sci.* **2015**, *112–113*, 75–93. [CrossRef]
40. Carazo, E.; Borrego-Sánchez, A.; García-Villén, F.; Sánchez-Espejo, R.; Aguzzi, C.; Viseras, C.; Sainz-Díaz, C.I.; Cerezo, P. Assessment of Halloysite Nanotubes as Vehicles of Isoniazid. *Colloids Surf. B Biointerfaces* **2017**, *160*, 337–344. [CrossRef]
41. Alkatheeri, M.S.; Palasuk, J.; Eckert, G.J.; Platt, J.A.; Bottino, M.C. Halloysite Nanotube Incorporation into Adhesive Systems—Effect on Bond Strength to Human Dentin. *Clin. Oral Investig.* **2015**, *19*, 1905–1912. [CrossRef]
42. Salmerón-Valdés, E.N.; Cruz-Mondragón, A.C.; Toral-Rizo, V.H.; Jiménez-Rojas, L.V.; Correa-Prado, R.; Lara-Carrillo, E.; Morales-Valenzuela, A.A.; Scougall-Vilchis, R.J.; López-Flores, A.I.; Hoz-Rodriguez, L.; et al. Mechanical Properties and Antibacterial Effect on Mono-Strain of Streptococcus Mutans of Orthodontic Cements Reinforced with Chlorhexidine-Modified Nanotubes. *Nanomaterials* **2022**, *12*, 2891. [CrossRef]
43. Takahashi, Y.; Imazato, S.; Kaneshiro, A.V.; Ebisu, S.; Frencken, J.E.; Tay, F.R. Antibacterial Effects and Physical Properties of Glass-Ionomer Cements Containing Chlorhexidine for the ART Approach. *Dent. Mater.* **2006**, *22*, 647–652. [CrossRef]
44. Hoszek, A.; Ericson, D. In Vitro Fluoride Release and the Antibacterial Effect of Glass Lonomers Containing Chlorhexidine Gluconate. *Oper. Dent.* **2008**, *33*, 696–701. [CrossRef]
45. Türkün, L.Ş.; Türkün, M.; Ertuğrul, F.; Ateş, M.; Brugger, S. Long-Term Antibacterial Effects and Physical Properties of a Chlorhexidine-Containing Glass Ionomer Cement. *J. Esthet. Restor. Dent.* **2008**, *20*, 29–44. [CrossRef]
46. Balouiri, M.; Sadiki, M.; Ibnsouda, S.K. Methods for in Vitro Evaluating Antimicrobial Activity: A Review. *J. Pharm. Anal.* **2016**, *6*, 71–79. [CrossRef]
47. Vermeersch, G.; Leloup, G.; Delmée, M.; Vreven, J. Antibacterial Activity of Glass-Ionomer Cements, Compomers and Resin Composites: Relationship between Acidity and Material Setting Phase. *J. Oral Rehabil.* **2005**, *32*, 368–374. [CrossRef]
48. Palenik, C.J.; Behnen, M.J.; Setcos, J.C.; Miller, C.H. Inhibition of Microbial Adherence and Growth by Various Glass Ionomers in Vitro. *Dent. Mater.* **1992**, *8*, 16–20. [CrossRef]
49. Bowden, G.H.W. Effects of Fluoride on the Microbial Ecology of Dental Plaque. *J. Dent. Res.* **1990**, *69*, 653–659. [CrossRef] [PubMed]
50. Perales Zamora, S.; Guillen Bordaz, C.; Loayza De La Cruz, R.; Alvarado Menacho, S.; Torres Ramos, G.; Guillén Astete, A.; Anticona Huaynate, C. El Flúor En La Prevención de Caries En La Dentición Temporal. Barnices Fluorados. *Odontol. Sanmarquina* **2014**, *9*, 33. [CrossRef]
51. Vamsi, K.; Siddiqui, F. Antimicrobial Effect of an Experimental Glass Ionomer Cement against Pathogens Associated with Deep Carious Lesions. *J. Contemp. Dent. Pract.* **2018**, *19*, 824–829. [CrossRef] [PubMed]
52. Heidarinasab, A.; Ahmad Panahi, H.; Faramarzi, M.; Farjadian, F. Synthesis of Thermosensitive Magnetic Nanocarrier for Controlled Sorafenib Delivery. *Mater. Sci. Eng. C* **2016**, *67*, 42–50. [CrossRef]
53. Renné, W.G.; Lindner, A.; Mennito, A.S.; Agee, K.A.; Pashley, D.H.; Willett, D.; Sentelle, D.; Defee, M.; Schmidt, M.; Sabatini, C. Antibacterial Properties of Copper Iodide-Doped Glass Ionomer-Based Materials and Effect of Copper Iodide Nanoparticles on Collagen Degradation. *Clin. Oral Investig.* **2017**, *21*, 369–379. [CrossRef] [PubMed]
54. Feitosa, S.A.; Palasuk, J.; Geraldeli, S.; Windsor, L.J.; Bottino, M.C. Physicochemical and Biological Properties of Novel Chlorhexidine-Loaded Nanotube-Modified Dentin Adhesive. *J. Biomed. Mater. Res. B Appl. Biomater.* **2019**, *107*, 868–875. [CrossRef]
55. Huang, Q.; Wang, S.; Sun, Y.; Shi, C.; Yang, H.; Lu, Z. Effects of Ag/ZnO Nanocomposite at Sub-Minimum Inhibitory Concentrations on Virulence Factors of Streptococcus Mutans. *Arch. Oral Biol.* **2020**, *111*, 104640. [CrossRef]
56. Feitosa, S.A.; Palasuk, J.; Kamocki, K.; Geraldeli, S.; Gregory, R.L.; Platt, J.A.; Windsor, L.J.; Bottino, M.C. Doxycycline-Encapsulated Nanotube-Modified Dentin Adhesives. *J. Dent. Res.* **2014**, *93*, 1270–1276. [CrossRef]
57. Massaro, M.; Lazzara, G.; Milioto, S.; Noto, R.; Riela, S. Correction: Covalently Modified Halloysite Clay Nanotubes: Synthesis, Properties, Biological and Medical Applications. *J. Mater. Chem. B* **2017**, *5*, 4246. [CrossRef]
58. Leyhausen, G.; Abtahi, M.; Karbakhsch, M.; Sapotnick, A.; Geurtsen, W. Biocompatibility of Various Light-Curing and One Conventional Glass-Ionomer Cement. *Biomaterials* **1998**, *19*, 559–564. [CrossRef] [PubMed]

59. Salmerón-Valdés, E.; Scougall-Vilchis, R.; Alanis-Tavira, J.; Morales-Luckie, R. Comparative Study of Fluoride Released and Recharged from Conventional Pit and Fissure Sealants versus Surface Prereacted Glass Ionomer Technology. *J. Conserv. Dent.* **2016**, *19*, 41–45. [CrossRef]
60. Kalagi, S.; Feitosa, S.A.; Münchow, E.A.; Martins, V.M.; Karczewski, A.E.; Cook, N.B.; Diefenderfer, K.; Eckert, G.J.; Geraldeli, S.; Bottino, M.C. Chlorhexidine-Modified Nanotubes and Their Effects on the Polymerization and Bonding Performance of a Dental Adhesive. *Dent. Mater.* **2020**, *36*, 687–697. [CrossRef]
61. Degrazia, F.W.; Leitune, V.C.B.; Takimi, A.S.; Collares, F.M.; Sauro, S. Physicochemical and Bioactive Properties of Innovative Resin-Based Materials Containing Functional Halloysite-Nanotubes Fillers. *Dent. Mater.* **2016**, *32*, 1133–1143. [CrossRef] [PubMed]
62. *CLSI M100;* Performance Standards for Antimicrobial Susceptibility Testing. Clinical and Laboratory Standards Institute CLSI: Berwyn, PA, USA, 2020.

inorganics

MDPI

Article

Nanocrystalline–Amorphous Transition in ZrN Nanofilms Induced by Helium Accumulation at Grain Boundaries

Xin Xiao [1,†], Sen Sun [2,†], Wei Jiang [3,*], Xiaoling Qin [4], Qinxin Liu [2] and Yuanxia Lao [4,*]

1 Department of Scientific Research, The People's Hospital of Guangxi Zhuang Autonomous Region, Nanning 530021, China; xiaoxi3891@163.com
2 School of Mathematics and Physics, Mianyang Teachers' College, Mianyang 621000, China; sensun@mtc.edu.cn (S.S.); tomasrao@outlook.com (Q.L.)
3 School of Mechanical Engineering, Hefei University of Technology, Hefei 230009, China
4 Guangxi Key Laboratory of Natural Polymer Chemistry and Physics, College of Chemistry and Materials, Nanning Normal University, Nanning 530001, China; qinxiaoling_00@163.com
* Correspondence: ljjw737@126.com (W.J.); 18656081809@163.com (Y.L.)
† These authors contributed equally to this work.

Abstract: Helium (He) accumulation, a byproduct of nuclear transmutation, poses a significant reliability challenge for the materials used in nuclear reactors. Nanomaterials, with their high density of interfaces, offer superior He tolerance by absorbing He atoms and suppressing bubble growth. However, the long-term stability of these materials under continuous He accumulation remains a concern. This study investigated the microstructural and mechanical property responses of ZrN nanofilms to excessive He accumulation. Different doses of He atoms were introduced via magnetron sputtering. The results indicate that increasing the He dose induced a nanocrystalline-to-amorphous transition and instability in the mechanical properties. The structural and mechanical instability, characterized by surface blistering, softening, abnormal lattice shrinkage, and amorphization, was primarily triggered by the degradation of the grain boundaries with He accumulation, and an amorphization model of nanomaterials is proposed.

Keywords: ZrN nanofilms; helium accumulation; grain boundaries; amorphization

1. Introduction

Under high-energy and high-flux neutron irradiation, reactor materials often undergo severe amorphization, leading to comprehensive performance degradation. To improve the radiation tolerance of nuclear materials, the idea of radiation-resistant nanomaterials has been proposed in recent years [1–4]. It uses high-density interfaces to inhibit the aggregation of radiation-induced point defects and accelerate their annihilation [5]. For example, oxide dispersion-strengthened (ODS) steels [6], multilayered materials [7], high-entropy alloys with nanoprecipitates [8], etc., have been reported that can annihilate defects and maintain a stable structure under radiation.

However, most of the research on radiation-tolerant nanomaterials has focused more on the damage caused by the atomic collision cascade of high-energy particles. In fact, as the energy and fluence of neutrons increase, the accumulation damage of transmutational products like He nuclei cannot be ignored [9,10]. The accumulation of He atoms can lead to the severe destruction of the structure and performance of nuclear materials [11]. Specifically, He atoms can trap themselves in vacancies to form He clusters (He bubbles), which introduce compressive stress into solids. With the bubbles' growth, cracks could form and evolve into cavities gradually, resulting in material embrittlement [12], surface blistering [13], etc., thus reducing the long-term reliability of nuclear materials.

Although some studies have shown that high-density grain boundaries in nanomaterials can significantly adsorb He bubbles and inhibit their growth in grains, giving nanomaterials superior tolerance to He accumulation damage [1–4], other researchers have reported some opposite conclusions. For example, when implanted with 10 keV He^+ ions at an average flux of 10^{13} He^+/cm^2s, large He-filled cavities can be formed in the grains of nanocrystalline nickel, facilitated by the radiation-induced vacancy clusters [14]. He accumulation can lead to surface blistering on the surface of nanocrystalline ZrN after He implantation and annealing at 600~1000 °C [15]. Chen et al. reported that with the implantation of 65 keV He^+ ions at 277 °C, and subsequent irradiation with 9 MeV Au ions (10 dpa) at 700 °C, He bubbles expanded obviously in nanocrystalline SiC, while no bubbles were found in a single-crystal SiC under the same irradiation conditions [16]. These findings suggest that the stability of nanomaterials under different He accumulation conditions needs to be evaluated carefully.

The stability of nanomaterials under He accumulation largely depends on the interaction behavior between the He atoms and grain boundaries [1–4]. However, the prevailing method to evaluate He accumulation damage is He^+ ion irradiation. Since ion irradiation will introduce inhomogeneous defects, thermal effects, and surface effects (limited depth of implantation) into the samples, it brings more challenges to quantitative studies on the pure interaction of He atoms and interfaces in nanomaterials, and increases the complexity of the failure behavior of nanomaterials. To investigate the pure interaction of grain boundaries and He atoms in nanomaterials, we tried to exclude the influence of inhomogeneous radiation defects and temperature and surface effects by introducing different doses of He atoms uniformly into ZrN nanofilms during film deposition, by magnetron sputtering at room temperature [17]. With this method, we evaluated the stability of ZrN nanofilms under severe He accumulation in this study, as ZrN is a promising radiation-tolerant material, especially as a candidate material for nuclear fuel cladding and other reactor structural components [18,19]. We demonstrate that He accumulation degrades the grain boundaries in ZrN nanofilms, driving amorphization and performance deterioration.

2. Results

2.1. Microstructure Evolution with He Accumulation

The typical GIXRD spectra of the ZrN nanofilms, obtained with He flow rates ranging from 0 to 70 standard cubic centimeters per minute (sccm), are shown in Figure 1. It was found that as the He flow rate increased, the full width at the half maximum of the peak became wider. Moreover, surface blistering on the specimen has become increasingly apparent with the He flow rate increasing (Figure 2). From the TEM images (Figure 3), it can be seen that the crystallite size within the ZrN thin films did not change significantly with the He flow rates, indicating the XRD peak broadening was substantially influenced by factors beyond the crystallite size, most likely by the non-uniform microstrain and decreased crystallinity induced by the He implantation. At an He flow rate of 45 sccm, the diffraction peaks broadened dramatically, showing that the crystallinity of the ZrN nanofilms decreased obviously. In particular, as the He flow rate increased to 70 sccm, all the peaks disappeared, indicating severe amorphization in the ZrN nanofilms. The peak positions of the different planes in the samples were obtained and converted to lattice parameters by using the software Jade 6.0. As shown in the inserted picture in Figure 1, the lattice parameters of the ZrN nanofilms with an He flow rate of 15 sccm increased slightly compared with the original ZrN. Then, it reduced sharply, from 4.63 Å to 4.52 Å, at an He flow rate of 45 sccm. The obvious shift of the (111) peak for the sample (45 sccm) toward a higher angle (marked by a dashed line) also verifies the lattice shrinkage. The abnormal lattice shrinkage followed by amorphization implies that the He flow rate of 45 sccm may be the threshold at which severe He cumulative damage occurred in the ZrN nanofilms under our synthesis conditions.

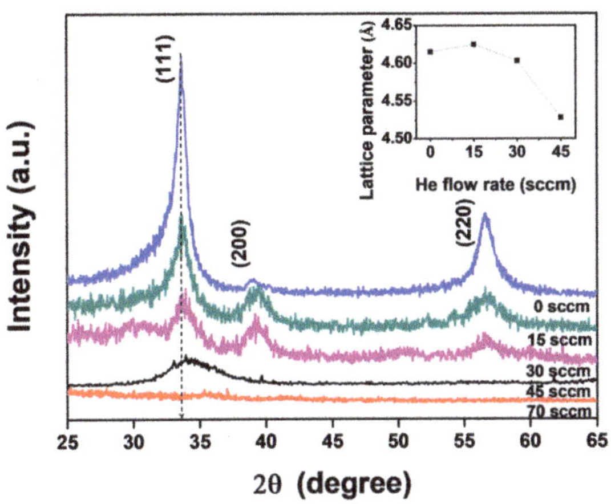

Figure 1. XRD patterns of ZrN nanofilms; the upper-right inset figure shows the change in the lattice constant with the He influx rate. "sccm" is the abbreviation of standard cubic centimeter per minute.

It is worth noting that many studies have shown that He accumulation often leads to lattice swelling, which can be attributed to the interstitial He atoms squeezing the lattice [11,18]. However, there was no monotonic increasing relationship between the lattice constant and the He dose in this study or in some other reports [17–21]. We will discuss this later, combining it with other findings.

2.2. Blistering on the Surface of ZrN Nanofilms

Prevalent blistering was found on the film surfaces at He flow rates above 45 sccm (Figure 2d,e), while the ZrN nanofilms had relatively flat and smooth surfaces (Figure 2a–c) below the threshold of 45 sccm. As we know, blistering on the surface is a typical pattern of He accumulation damage, resulting from the release of He atoms and other interstitial atoms diffusing from the interior of the sample [17]. Lattice shrinkage and blistering happening at the same time implies a close relationship between these two damage behaviors. It is reasonable to suspect that noticeable lattice shrinkage can predict a qualitative change in the microstructure of nanofilms, which may trigger severe He accumulation damage.

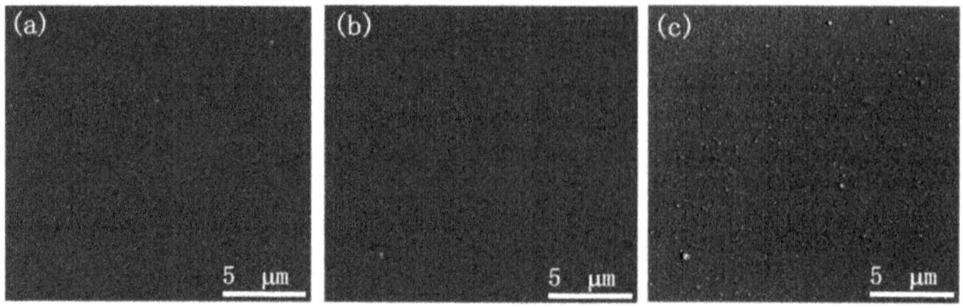

Figure 2. *Cont.*

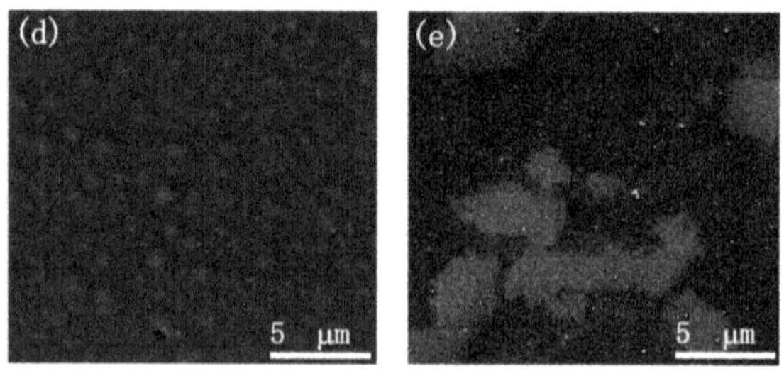

Figure 2. Surface morphology of the (**a**) as-deposited ZrN nanofilm and (**b**) He-implanted ZrN nanofilm at 15 sccm, (**c**) 30 sccm, (**d**) 45 sccm, and (**e**) 70 sccm, showing that blisters appear as the He flow rate increases.

2.3. Interaction of He Bubbles and Grain Boundaries

To understand the microstructure evolution behavior as the He injection rate increased, we compared the cross-sectional TEM images of the samples with different He doses. Figure 3a shows the TEM image of the nanofilm without He injection. It can be seen that the columnar crystal structure inside the film is relatively intact. Moreover, the four sharp—(111), (200), (220), and (311)—diffraction rings in the selected area electron diffraction (SAED) pattern (inset of Figure 3a) indicate that the ZrN nanofilm without He injection has better crystallinity. The columnar grains in the ZrN films were interrupted at the He flow rate of 15 sccm, resulting in more equiaxed grains of 10–20 nm in diameter, as shown in Figure 3b. Moreover, the initial He bubbles were mainly adsorbed at the grain boundaries in the form of strips (marked by white arrows). The strip-shaped He bubbles began to coarsen as the He dose increased (Figure 3c), but most of the He bubbles were still located along the grain boundaries. At the He flow rate of 45 sccm, a large number of irregular polygonal He bubbles can be observed in Figure 3d. Meanwhile, the grain boundaries became blurred and the He bubbles expanded from the grain boundaries toward the interior of the grain. We hypothesize that the irregular polygonal He bubbles formed when the bubbles, initially at the grain boundaries, began expanding into the grain interior as the grain boundary structure was compromised. We suggest that a residual pinning effect from the original grain boundary location, influenced by the interfacial energy, prevented the expanding bubble edges from fully relaxing into a spherical shape. The diffraction ring became blurred, indicating that its crystallinity had decreased severely (Figure 3d). As the He flow rate increased to 70 sccm, the He bubbles aggregated and expanded dramatically, while the grain boundaries disappeared and the irregular polygonal He bubbles became spherical (Figure 3e). The bubble shape changing indicates that the He bubbles were completely free from the grain boundaries and evolved to a spherical shape driven by the Gibbs free energy. This observation is supported by the diffraction rings (inset in Figure 3e). They transformed into halos, revealing that the nanocrystals underwent severe amorphization. This indicates that with increases in the He dose, the trapping (accommodating) effect of the grain boundaries on the He bubbles gradually weakens and then disappears, resulting in remarkable structural damage to the nanofilms.

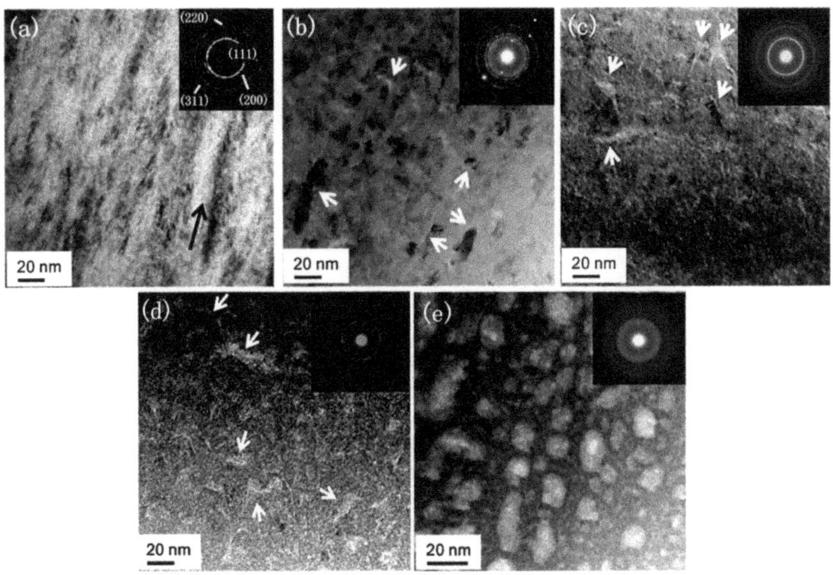

Figure 3. Cross-sectional TEM images of the ZrN nanofilms: (**a**) the as-deposited ZrN nanofilm, (**b**) He-implanted ZrN nanofilm with 15 sccm, (**c**) He-implanted ZrN nanofilm with 30 sccm, (**d**) He-implanted ZrN nanofilm with 45 sccm, and (**e**) He-implanted ZrN nanofilm with 70 sccm. The black arrow shows the growth direction of the grain. The white arrows mark the He bubbles. The upper-right inset figures show the corresponding electron diffraction patterns.

The high-resolution transmission electron microscope (HRTEM) images reveal the interaction between the He atoms and grain boundaries. Figure 4a shows the intact grain boundaries of the nanofilms without He bubbles, while Figure 4b shows stripe-shaped He bubbles (indicated by the white arrows) distributed along the grain boundaries and a small amount of He bubbles (indicated by the black arrows) in the grain interior, which confirms the trapping effect of the grain boundary on the He bubbles, keeping the crystal structure stable at the He flow rate of 15 sccm. With an increase in the He dose, the strip-shaped He bubbles on the grain boundary began to gather and coarsen (Figure 4c). Note that the expansion of the spherical bubbles in the grain (indicated by the black arrows) is not obvious. This reveals that the growth of the He bubbles was still confined within the grain boundary at the He flow rate of 30 sccm. However, when the He flow rate increased to 45 sccm, as shown in Figure 4d, most of the stripe-shaped He bubbles along the grain boundaries grew into irregular polygonal bubbles (marked by the white arrows). And the bubbles expanded significantly toward the interior of the grains and swallowed up the bubbles in the grains forming larger He bubbles (the white dotted line). This shows the weak influence of the grain boundary on the He bubbles once the He dose surpasses the threshold. Finally, as the He dose increased, the grain boundary disappeared and the lattice structure was completely destroyed, showing an amorphous state (Figure 4e). Without the limitation of the grain boundary, the He bubbles spontaneously converged into a spherical shape to reduce the surface energy. The interaction behavior of the He bubbles and the grain boundary in Figure 4 is consistent with the experimental results shown in Figures 1–3. It is rational to suspect that the weakening of the trapping effect of the grain boundary on the He bubbles was the essential factor leading to the severe He accumulation damage, including lattice shrinkage, amorphization, and blistering. And the He dose surpassing the threshold may have been the trigger for the degradation of the grain boundary.

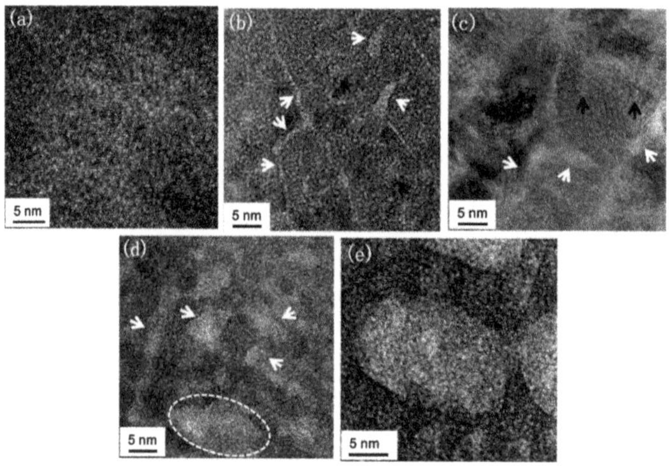

Figure 4. High-resolution transmission electron microscope (HRTEM) images of the ZrN nanofilms; the corresponding flow rates are (**a**) 0 sccm, (**b**) 15 sccm, (**c**) 30 sccm, (**d**) 45 sccm, and (**e**) 70 sccm, respectively. The white arrows mark the He bubbles at or near the grain boundaries, while the black arrows mark the He bubbles in the grain interior. The white circle marks the expansion of the He bubbles.

2.4. Degradation of Grain Boundaries Induced by He Accumulation

To reveal the degradation mechanism of the grain boundary, we further analyzed the microstructure of typical grains at different He doses through HRTEM. Figure 5 presents the fast Fourier transform (FFT) and inverse Fourier transform (IFFT) of the HRTEM images within a selected area (within the dotted box). Figure 5a shows the HRTEM image of the film with the He flow rate of 15 sccm. At this low He dose, more He bubbles were located at the grain boundaries, with a few small bubbles in the grain interior (see the dotted box). From the FFT in the upper-left corner of Figure 5a, it can be seen that the diffraction spots in the crystal planes of the selected grain remain intact and regular. Furthermore, as shown in the IFFT (Figure 5b), the interplanar distance of the {111} plane is 2.67 Å in the grain, meaning the lattice constant of the selected grain is a little higher than that of the sample without He injection. The lattice swelling can be attributed to the small He bubbles squeezing the lattice in the grain (marked by the red arrows). However, the lattice swelling is limited and the lattice structure is intact because most of the He bubbles have been trapped in the grain boundaries. It is believed that the trapping effect of the grain boundary on He atoms can improve the resistance to He accumulation damage in ZrN nanofilms.

When the He flow rate was 45 sccm, the lattice changed greatly. The He bubbles broke free from the grain boundaries and expanded into the interior of the grain (the dotted box in Figure 5c). In the FFT corresponding to Figure 5c, the reciprocal lattice corresponding to the crystal plane is blurred, and satellite points (marked by the red arrows) appear. The two satellite points lie in the line of the reciprocal points of the {111} plane, but drift outward for a small distance. This indicates that part of the {111} plane collapsed, and lattice shrinkage occurred. The IFFT in Figure 5d shows that the collapse of the {111} planes occurred from the zone near the grain boundaries (marked by the red arrows) and expanded toward the grain interior, resulting in a decrease in the interplanar distance of the {111} plane to 2.58 Å.

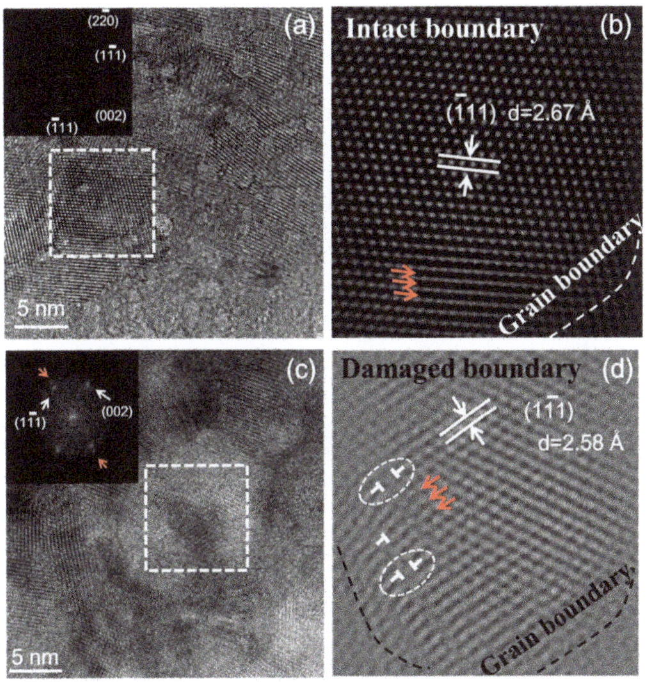

Figure 5. HRTEM images and fast Fourier transform (FFT) for the rectangle areas: (**a**) 15 sccm He injection sample and (**c**) 45 sccm He injection sample. Corresponding inverse Fourier transform (IFFT) of the rectangle areas: (**b**) 15 sccm He injection sample and (**d**) 45 sccm He injection sample. Grain boundary degradation appears with an He dose surpassing the threshold, being distinguished by the prevalence of dislocations near the grain boundaries. The white arrow and red arrow point out the intact crystal plane and the deformed crystal plane, respectively.

It is important to note that a large number of vacancy-type defects (Figure 5d), like dislocations (marked by the Ts) and dislocation loops (marked by the dotted lines), appeared near the grain boundaries (periphery zone in Figure 5d), which were rarely found in the grain with a low He dose (Figure 5b). We believe it was the prevalence of the vacancy-type defects that destroyed the structure of the grain boundary, and weakened its trapping effect on the He bubbles. Plenty of the vacancy-type defects near the grain boundaries can be explained by the dislocation loop-punching mechanism [22]. At low He doses, more He-vacancy clusters preferentially accumulated within the grain boundaries because the energy barriers for point defect migration were lower at the interfaces than in the bulk [23]. And the growth of the He bubbles could be confined within the grain boundaries because the intact lattice structure near the interface inhibited the diffusion of the point defects. This resulted in more strip-shaped He bubbles growing along the grain boundaries and a few spherical He bubbles forming in the grain interior. As the He dose increased, no more space in the grain boundary was left for the He bubbles. Then, the strip-shaped bubbles tended to expand toward the grain interior by punching out dislocation loops constantly, leading to more vacancy-type defects forming near the interfaces. As the He dose increased to the threshold, the cumulative vacancy-type defects destroyed the lattice structure near the grain boundaries (Figure 5d), which reduced the ability of the interfaces to confine the He bubbles.

As the He bubbles expanded toward the grain interior, they merged with the bubbles in the grain (Figures 4d and 5c). This process led to the release of the cumulative compressive stress induced by the He bubbles, which decreased the lattice constant [17–21]. Moreover, the destruction of the interface induced by the prevailing vacancy-type defects near the grain boundaries enhanced the short-circuiting diffusion. More interstitial atoms (like He atoms and self-interstitial atoms) tended to migrate along the grain boundaries toward the surface to further release the compressive stresses caused by the high-dose He injection. The migration of these massive interstitial atoms to the surface led to the lattice collapse (lattice shrinkage) and blistering of the ZrN nanofilms.

It is worthy to note that the He-induced lattice changes can also be readily interpreted from the perspective of macroscopic film stress evolution. At low He doses, the observed increase in the ZrN lattice parameter suggests the incorporation of interstitial He atoms, likely inducing compressive stress. Conversely, at higher He doses, pronounced grain boundary damage facilitated interstitial atoms diffusion and surface blistering. This process is typically associated with the formation of vacancy-type defects within a lattice, leading to a tendency toward tensile stress and observed lattice contraction.

2.5. Mechanical Properties of ZrN Nanofilms

As shown in Figure 6, the indentation hardness of the film without He injection is about 19 (Gpa), and the elastic modulus is about 185 (Gpa). The hardness and elastic modulus of the films with He are lower than those of the original one, and both of them decreased as the He dose increased. Especially when the He flow rate reached 45 sccm, both the hardness and elastic modulus dropped sharply. This is consistent with our supposition that the He flow rate of 45 sccm was the threshold for He accumulation damage in the films under our synthesis conditions.

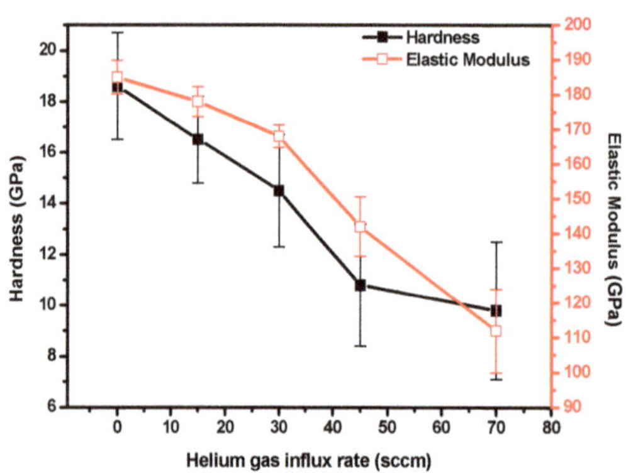

Figure 6. Hardness and elastic modulus decreases with the He dose.

The hardness deteriorated with an increase in the He dose in this study, which is contradictory to some previous studies. Many researchers have reported that the hardness of He-injected samples is positively correlated with the dose of the He injection [24]. This has been attributed to the defects introduced by the He bubbles hindering the movement of dislocations. However, both the hardness and elastic modulus of the ZrN nanofilms were inversely correlated with the He dose in this study, implying that dislocation movement may be no longer the main deformation mode of He-injected materials, which will be explained in the Section 3.

3. Discussion

Researchers often believe that high-density grain boundaries can improve the resistance to He accumulation damage. However, this study shows that as the He dose surpassed the threshold, the grain boundaries were preferentially destroyed, leading to severe amorphization and performance deterioration. Here, we propose a model for the various stages of He accumulation damage in nanomaterials, as shown in Figure 7. In the first stage, He accumulation in the nanomaterials is limited. The high density of the grain boundaries in the nanomaterials can adsorb the He atoms and inhibit the expansion of He bubbles toward the grains, thus improving the material's resistance to He accumulation damage (Figure 7a). However, as the He accumulation increases to the threshold, the lattice structure of the grain boundaries and their neighboring zones are destroyed, resulting from the He bubbles expanding toward the grain interior. During this stage, the expansion of the He bubbles introduces more self-interstitial atoms near the grain boundaries, which diffuse along the grain boundaries to the surface, driven by the surface energy (Figure 7b). This leads to lattice shrinkage in the nanocrystalline structure and blistering on the surface, resulting in the deterioration of the mechanical performance. As the He accumulation increases continuously, the structure of the grain boundaries is severely destroyed. Out of the control of the grain boundaries, the He bubbles expand deeply into the grains, merging and growing dramatically. At the end of this stage, a huge amount of self-interstitial atoms are introduced by the expansion of the He bubbles, which can lead to the amorphization of the nanomaterials (Figure 7c).

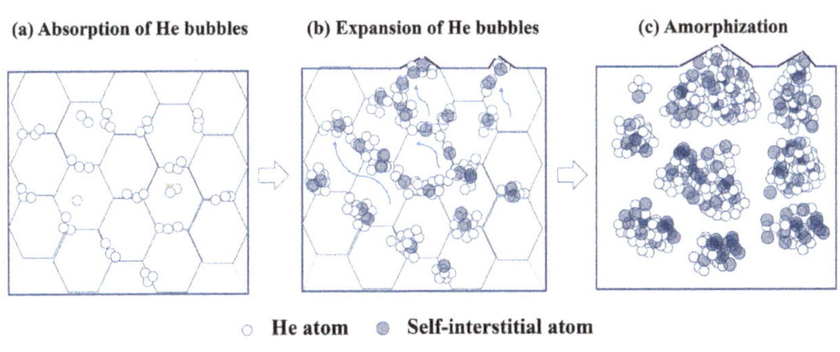

Figure 7. Amorphization model of nanomaterials with He accumulation.

This model can explain the degradation behavior of the hardness and elastic modulus of ZrN nanofilms with He accumulation (Figure 6). It has been reported that the deformation of nanomaterials can be more affected by the grain sliding, rotation, and dislocation emissions of the grain boundaries [25–27]. In this study, the grain boundaries in the ZrN nanofilms were destroyed by the prevailing vacancy-type defects (like dislocation loops) induced by the He accumulation. The degradation of the grain boundaries enhanced the sliding and rotation of the grains, and more dislocations were emitted from the grain boundaries (Figure 5d), resulting in a decrease in hardness. For the elastic modulus, the degradation was mainly related to the chemical bonds and crystal structure. The vacancy-type defects induced by the excessive He atoms weakened the chemical bonding force and destroyed the lattice structure near the grain boundaries. This caused a decrease in the Peierls stress in the nanofilms, resulting in the decrease in the elastic modulus.

The above hypothesis regarding the mechanism of the mechanical property degradation is consistent with the typical damage behavior we observed in the high-He-dose films (see Figure S3 in the Supplementary Materials). Under the He flow rate of 45 sccm, some cavities were found beneath the surface blisters (Figure S3a by SEM). The subsurface layer's cross-sectional profile may show the origin of the blistering (Figure S3b by TEM). Some cracks appeared in the subsurface layer, mainly propagated along the grain boundaries. It is rational to believe that the degradation of the grain boundaries accelerated crack initiation and propagation along the interfaces, and then enhanced the diffusion of the interstitial atoms (He and self-interstitial atoms) to the surface, forming cavities and blisters. The formation of blisters, cavities, and cracks, accelerated by the degradation of the grain boundaries, degraded the mechanical properties. This also accounts for the observed reduction in the strength and elastic modulus of the nanomaterial resulting from the substantial He accumulation damage in this study.

4. Experimental Section

The total thickness of all the He-implanted films was kept at about 1 μm. For the film deposition, a pure zirconium plate (99.995%, Grinm Advanced Materials Co., Ltd., Beijing, China) was used as the target, with a size of 65 mm in diameter and 5 mm in thickness. The target–substrate distance was 50 mm. The film deposition was conducted in a magnetron sputtering chamber, which was evacuated to 1.0×10^{-4} Pa. Before deposition, N-type single-crystal Si (100) single-sided polished sheets (Grinm Advanced Materials Co., Ltd., Beijing, China), used as the substrates, were continuously cleaned with ethanol, acetone, and deionized water. The substrates in the chamber were cleaned again by applying RF power of 200 W under Ar (99.999%, Zhongke Gas Co., Ltd., Nanning, China) for 10 min. During the film deposition, pure Ar and N_2 (99.999%, Zhongke Gas Co., Ltd., Nanning, China) were introduced into the chamber with influx rates of 45 sccm (standard cubic centimeter per minute) and 25 sccm, and pure He gas (99.999%, Zhongke Gas Co., Ltd., Nanning, China) was injected into the chamber with influx rates of 0, 15, 30, and 45 sccm, respectively. The sputtering power was 100 W, and the substrate holder was neither cooled nor heated during sputtering. ZrN nanofilms were obtained after 2 h of deposition, and their thickness was about 1 μm. It is worth noting that He can be uniformly introduced into ZrN nanofilms using the magnetron sputtering method, and the introduced He concentration can be controlled via the He influx rates. Analysis of TEM bright-field images (Figure S1) combined with ImageJ software-v1.8.0 processing (Figure S2) revealed that the total surface area of He bubbles, encompassing various morphologies, increased with the He influx rates. This indicates that qualitatively controlling the He implantation dose through the He influx rates is feasible.

The crystal structure of the films was characterized by glancing-incident-angle X-ray diffractometer (GIXRD, Philips X Pert Pro MPD DY129, Koninklijke Philips N.V., Amsterdam, The Netherlands), and the changes in lattice structure before and after He injection were observed. The glancing angle was 1°, with a scanning speed of 5°/min. Field-emission scanning electron microscopy (SEM, Hitachi SU 8000, Hitachi, Ltd., Tokyo, Japan) was used to characterize the surface and cross-sectional morphology of the films, observing the blisters or cracks on the films. The acceleration voltage was 15 kV. Transmission electron microscope (TEM, Tecnai G20 S-T win, Fei Company, Hillsboro, OR, USA) was used to characterize the microstructure evolution with different He doses (changed by flow rates), including lattice distortion, bubble growth, and defect pattern in the ZrN nanofilms.

The effect of He accumulation on the mechanical properties of ZrN nanofilms was determined by using a nano-indentator (Nano Indenter G200, KLA Corporation, Milpitas, CA, USA). The nanoindentation measurement was conducted at a temperature of 23 °C, using a Berkovich indenter. To minimize the impact of the substrate on the film hardness, the applied maximum load was set at 6 mN, which made sure the indentation depth did not surpass 20% of the thickness of the films. Each sample was subjected to 8 independent indentation tests to reduce the experimental error.

5. Conclusions

Excessive He accumulation can lead to the amorphization and performance degradation of ZrN nanofilms. The structural integrity of the grain boundary plays an essential role in the nanocrystalline–amorphous transition. At low He doses, the high-density grain boundaries can absorb He atoms, thus suppressing the lattice damage induced by He bubbles. However, as the He dose increases, the grain boundary can be destroyed by the He accumulation. The degradation of the grain boundaries triggers the He bubbles' expansion toward the grain interior, leading to structural and mechanical instability, including surface blistering, softening, lattice shrinkage, and amorphization.

This finding gives us a more comprehensive view of the interaction between grain boundaries and He bubbles, and the amorphization of ZrN nanofilms with He accumulation. It provides an essential guide for the design of new irradiation-resistant nanomaterials.

Supplementary Materials: The following supporting information can be downloaded at: https://www.mdpi.com/article/10.3390/inorganics13050158/s1, Figure S1: Bright-field TEM micrographs of ZrN films with different He gas influx rates. Figure S2: The ratio of area of He bubbles in ZrN films at different He influx rates including spherical He bubbles, platelet-like He bubbles, and He bubbles in total. The analysis was conducted using ImageJ software on Figure S1. Figure S3: Cross-sectional morphology of the damaged ZrN nanofilms at a flow rate of 45sccm: (a) top surface layer captured by SEM, (b) subsurface layer captured from TEM. Arrows mark the defects near the surface of ZrN films.

Author Contributions: Conceptualization, X.X. and Y.L.; data collection, S.S. and W.J.; data interpretation, X.X., Q.L., S.S. and X.Q.; writing—original draft preparation, S.S.; writing—review and editing, X.Q. and X.X.; supervision, W.J. and Y.L. All authors have read and agreed to the published version of the manuscript.

Funding: This project was supported by the National Natural Science Foundation of China (Grant Nos. 11505121 and 12205134), the Guangxi Science and Technology Program (Grant No. AB24010253), the Guangxi Key Research and Development Program (No. Guike AB23026047), and the Key Teacher Training Program of Guangxi Province in 2021.

Institutional Review Board Statement: Not applicable.

Informed Consent Statement: Not applicable.

Data Availability Statement: The data are contained within the article.

Acknowledgments: Y. Lao wishes to express his sincere gratitude to M. Amy for sharing her inspiration to study.

Conflicts of Interest: The authors declare no conflicts of interest.

References

1. Yang, W.; Pang, J.; Zheng, S.; Wang, J.; Zhang, X.; Ma, X. Interface effects on He ion irradiation in nanostructured materials. *Materials* **2019**, *12*, 2639. [CrossRef] [PubMed]
2. Wang, H.; Araujo, R.; Swadener, J.G.; Wang, Y.Q.; Zhang, X.; Fu, E.G.; Cagin, T. Ion irradiation effects in nanocrystalline TiN coatings. *Nucl. Instrum. Methods Phys. Res. B* **2007**, *261*, 1162–1166. [CrossRef]

3. Jiao, L.; Chen, A.; Myers, M.T.; General, M.J.; Shao, L.; Zhang, X.; Wang, H. Enhanced ion irradiation tolerance properties in TiN/MgO nanolayer films. *J. Nucl. Mater.* **2013**, *434*, 217–222. [CrossRef]
4. Kim, I.; Jiao, L.; Khatkhatay, F.; Martin, M.; Lee, J.; Shao, L.; Zhang, X.; Swadener, J.; Wang, Y.; Gan, J.; et al. Size-dependent radiation tolerance in ion irradiated TiN/AlN nanolayer films. *J. Nucl. Mater.* **2013**, *441*, 47–53. [CrossRef]
5. Beyerlein, I.J.; Demkowicz, M.J.; Misra, A.; Uberuaga, B.P. Defect-interface interactions. *Prog. Mater. Sci.* **2015**, *74*, 125–210. [CrossRef]
6. Sagaradze, V.V.; Kozlov, K.A.; Kataeva, N.V. Oxide-Dispersion Strengthened Radiation-Resistant Steels. *Phys. Met. Metallogr.* **2018**, *119*, 1350–1353. [CrossRef]
7. Misra, A.; Demkowicz, M.J.; Zhang, X.; Hoagland, R.G. The radiation damage tolerance of ultra-high strength nanolayered composites. *JOM* **2007**, *59*, 62–65. [CrossRef]
8. Kombaiah, B.; Zhou, Y.; Jin, K.; Manzoor, A.; Poplawsky, J.D.; Aguiar, J.A.; Bei, H.; Aidhy, D.S.; Edmondson, P.D.; Zhang, Y. Nanoprecipitates to Enhance Radiation Tolerance in High-Entropy Alloys. *ACS Nano* **2023**, *15*, 3912–3924. [CrossRef]
9. Odette, G.R.; Alinger, M.J.; Wirth, B.D. Recent developments in irradiation-resistant Steels. *Annu. Rev. Mater. Res.* **2008**, *38*, 471–503. [CrossRef]
10. Zinkle, S.J.; Busby, J.T. Structural materials for fission & fusion energy. *Mater. Today* **2009**, *12*, 12–19.
11. Uglov, V.; Abadias, G.; Zlotski, S.; Kvasov, N.; Saladukhin, I.; Malashevih, A. Blister formation in ZrN/SiN multilayers after He irradiation. *Surf. Coat. Technol.* **2018**, *355*, 311–317. [CrossRef]
12. Lasa, A.; Tähtinen, S.K.; Nordlund, K. Loop punching and bubble rupture causing surface roughening—A model for W fuzz growth. *Europhys. Lett.* **2014**, *105*, 25002. [CrossRef]
13. Chen, C.; Zhang, Y.; Wang, Y.; Crespillo, M.; Fontana, C.; Graham, J.; Duscher, G.; Shannon, S.; Weber, W. Dose dependence of helium bubble formation in nano-engineered SiC at 700 °C. *J. Nucl. Mater.* **2016**, *472*, 153–160. [CrossRef]
14. Muntifering, B.; Blair, S.J.; Gong, C.; Dunn, A.; Dingreville, R.; Qu, J.; Hattar, K. Cavity evolution at grain boundaries as a function of radiation damage and thermal conditions in nanocrystalline nickel. *Mater. Res. Lett.* **2015**, *4*, 96–103. [CrossRef]
15. Van Vuuren, A.J.; Neethling, J.H.; Skuratov, V.A.; Uglov, V.V.; Petrovich, S. The effect of He and swift heavy ions on nanocrystalline zirconium nitride. *Nucl. Instrum. Methods Phys. Res. B* **2014**, *326*, 19–22. [CrossRef]
16. Chen, C.; Zhang, Y.; Fu, E.; Wang, Y.; Crespillo, M.; Liu, C.; Shannon, S.; Weber, W. Irradiation-induced microstructural change in helium-implanted single crystal and nano-engineered SiC. *J. Nucl. Mater.* **2014**, *453*, 280–286. [CrossRef]
17. Lao, Y.; Niu, W.; Shi, Y.; Du, H.; Zhang, H.; Hu, S.; Wang, Y. Effects of intragranular defects on helium accumulation damage in nanocrystalline nitrides. *J. Alloys Compd.* **2018**, *739*, 401–406. [CrossRef]
18. Kleykamp, H. Selection of materials as diluents for burning of plutonium fuels in nuclear reactors. *J. Nucl. Mater.* **1999**, *275*, 1–11. [CrossRef]
19. Wheeler, K.; Peralta, P.; Parra, M.; McClellan, K.; Dunwoody, J.; Egeland, G. Effect of sintering conditions on the microstructure and mechanical properties of ZrN as a surrogate for actinide nitride fuels. *J. Nucl. Mater.* **2007**, *366*, 306–316. [CrossRef]
20. Janse van Vuuren, A.; Skuratov, V.; Uglov, V.; Neethling, J.; Zlotski, S. Radiation tolerance of nanostructured ZrN coatings against swift heavy ion irradiation. *J. Nucl. Mater.* **2013**, *442*, 507–511. [CrossRef]
21. Taylor, C.A.; Patel, M.K.; Aguiar, J.A.; Zhang, Y.; Crespillo, M.L.; Wen, J.; Xue, H.; Wang, Y.; Weber, W.J. Bubble formation and lattice parameter changes resulting from He irradiation of defect-fluorite $Gd_2Zr_2O_7$. *Acta Mater.* **2016**, *115*, 115–122. [CrossRef]
22. Trinkaus, H.; Wolfer, W. Conditions for dislocation loop punching by helium bubbles. *J. Nucl. Mater.* **1984**, *122*, 552–557. [CrossRef]
23. Bai, X.M.; Voter, A.F.; Hoagland, R.G.; Nastasi, M.; Uberuaga, B.P. Efficient annealing of radiation damage near grain boundaries via interstitial emission. *Science* **2010**, *327*, 1631–1634. [CrossRef]
24. Wei, Y.; Liu, P.; Zhu, Y.; Wang, Z.; Wan, F.; Zhan, Q. Evaluation of irradiation hardening and microstructure evolution under the synergistic interaction of He and subsequent Fe ions irradiation in CLAM steel. *J. Alloys Compd.* **2016**, *676*, 481–488. [CrossRef]
25. Zhu, Y.T.; Langdon, T.G. Influence of grain size on deformation mechanisms: An extension to nanocrystalline materials. *Mater. Sci. Eng. A* **2005**, *409*, 234–242. [CrossRef]
26. Wang, L.; Zhang, Y.; Zeng, Z.; Zhou, H.; He, J.; Liu, P.; Chen, M.; Han, J.; Srolovitz, D.J.; Teng, J.; et al. Tracking the sliding of grain boundaries at the atomic scale. *Science* **2022**, *375*, 1261–1265. [CrossRef]
27. Hu, J.; Shi, Y.N.; Sauvage, X.; Sha, G.; Lu, K. Grain boundary stability governs hardening and softening in extremely fine nanograined metals. *Science* **2017**, *355*, 1292–1296. [CrossRef]

inorganics

MDPI

Article

Impact-Induced Plastic Deformation in CuZr Metallic Glass and MG/Cu Composites

Nicolás Amigo *, Javier Wachter and Pablo Leiva-Pavés

Departamento de Física, Facultad de Ciencias Naturales, Matemática y del Medio Ambiente, Universidad Tecnológica Metropolitana, Las Palmeras 3360, Ñuñoa 7800003, Chile

* Correspondence: namigo@utem.cl

Abstract: The mechanical response of monolithic CuZr metallic glass (MG) and MG/Cu composite substrates under high-velocity impact was investigated using molecular dynamics simulations, with variations in impact velocity and initial temperature. Higher impact velocities resulted in deeper penetration and increased plastic deformation, with the monolithic MG exhibiting greater energy absorption and slightly more extensive projectile fragmentation. The MG/Cu composite displayed enhanced plastic deformation, attributed to the higher stiffness of the crystalline Cu phase, which promoted plasticity in the amorphous matrix. Temperature effects were more pronounced in the composite, where elevated temperatures enhanced strain localization and atomic mobility in the glassy phase. This was supported by a decrease in dislocation density and the population of hexagonal close-packed (HCP) atoms with increasing temperature, indicating a shift in plastic activity toward the amorphous matrix. These findings provide insights into the interplay between impact velocity, temperature, and material composition, contributing to a deeper understanding of MG-based composite behavior under extreme loading conditions.

Keywords: impacts; plasticity; metallic glasses; composites; molecular dynamics

1. Introduction

The impacts of nanoparticles have garnered increasing attention due to their relevance in ballistics [1], the aerospace industry [2], military applications [3], and other fields [4,5]. High-velocity impacts involve various significant physical phenomena. For example, crater morphology offers valuable insights into energy dissipation and pressure distribution [6]. The deformation mechanism is highly dependent on the material's properties. Brittle materials, such as ceramics and glass, typically exhibit catastrophic failure, absorbing less energy through plastic deformation and forming craters characterized by sharp edges and radial cracks. In contrast, ductile materials undergo extensive plastic deformation, resulting in smoother and more rounded craters [7–10]. Moreover, mechanical properties such as hardness and stiffness play a crucial role in determining a material's resistance to localized deformation and penetration, significantly influencing its initial response under impact [11–13].

Understanding the underlying mechanisms of material damage requires high-precision techniques, which are often difficult to achieve experimentally. To address this limitation, numerical modeling has emerged as a powerful and cost-effective tool for studying high-velocity impacts with high accuracy [14,15]. The finite element method (FEM) has been widely used to investigate macroscopic systems and large structures. Applications include evaluating ballistic limit thickness for armor steels [16], analyzing the perforation of metallic plates [17], assessing the resistance of composite laminated plates [18], and studying traumatic brain injury [19], among others. Despite its versatility, FEM lacks atomic-level resolution and relies heavily on constitutive material models. In contrast, molecular dynamics (MD) provides atomic-scale detail, though it depends on the accuracy of the employed interatomic potentials. While MD has primarily been applied to thermal spray processes [20–23], other studies have investigated the effects of crystallographic orientation on deformation [24], the plastic response and surface damage in metallic glasses (MGs) [25,26], and wall cratering in tungsten surfaces [27]. However, aspects such as the mechanical response of composites and the influence of secondary phase distribution remain largely unexplored, presenting ongoing challenges in this field.

The mechanical behavior of MGs and MG-based composites under high-velocity impact is of particular interest due to their unique atomic structure and exceptional mechanical properties. MGs feature a disordered atomic arrangement that enhances their strength and elastic limit, while limiting their capacity for plastic deformation [28–30]. The introduction of a crystalline phase, as in MG/Cu composites, adds structural heterogeneity that can significantly influence energy dissipation and deformation mechanisms. However, the interaction between the amorphous matrix and the crystalline phase under extreme loading conditions remains poorly understood, particularly with respect to shear strain localization, dislocation activity, and atomic rearrangement [31–34]. Temperature also plays a critical role in the mechanical response during impact. Elevated temperatures can enhance atomic mobility and promote plastic deformation in the amorphous phase, while reducing dislocation density in the crystalline phase [35–37]. These effects are especially relevant for understanding the transition between brittle and ductile behavior, as well as the evolution of damage mechanisms at varying temperatures. Prior studies have also investigated the shock response of MGs, characterizing properties such as Hugoniot pressure and spall strength [38–41]. While these works advance the understanding of material response under extreme and transient loading, they do not adequately characterize the damage caused by localized impacts. This gap highlights the need for further investigation into high-velocity impacts to fully understand damage evolution in MGs and MG-based composites.

In this context, the present study investigates the mechanical response of monolithic CuZr MG and MG/Cu composite substrates under high-velocity impact using MD simulations. A nanoporous MG matrix was prepared, with a secondary Cu crystalline phase stochastically embedded to occupy the porous regions. By systematically varying the impact velocity and initial temperature, this work aims to elucidate the roles of structural heterogeneity and thermal effects on plastic deformation, energy dissipation, and shear strain localization, topics that have rarely been addressed in the literature, to the best of the authors' knowledge. This approach offers valuable insights into the behavior of MG-based composites under extreme loading conditions and contributes to the development of advanced materials for impact-resistant applications.

2. Results and Discussion

In this section, the response of the monolithic MG and the MG/Cu composite to nanoparticle impact is analyzed as a function of impact velocity and initial substrate temperature. The focus is on the structural and mechanical properties of the substrate once it reaches a relaxed state, which occurs approximately 0.6 ns after impact.

2.1. Deformation Fields

The deformation behavior of both the monolithic MG and the MG/Cu composite at an initial temperature of 100 K was analyzed using atomic shear strain for the amorphous phase and common neighbor analysis (CNA) for the crystalline phase, as shown in Figure 1. Higher impact velocities resulted in deeper penetration and increased projectile deformation, ultimately leading to disintegration at 3.0 km/s, consistent with previous studies on high-velocity impacts in metallic materials [26,42]. A comparison between the monolithic and composite samples reveals similar overall impact behavior. However, in the MG/Cu composite, plasticity is partially driven by deformation within the crystalline phase, resulting in the nucleation of stacking faults, a phenomenon typically observed in face-centered cubic (FCC) metals [43–45].

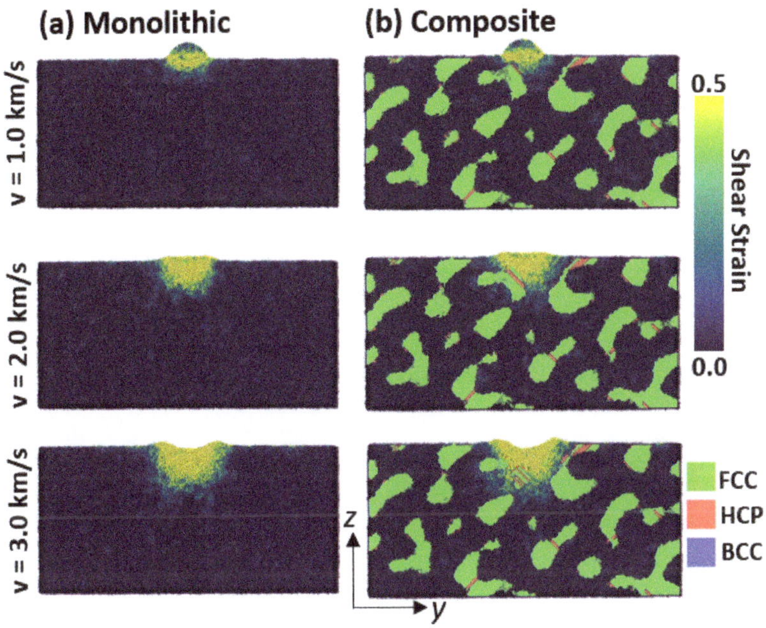

Figure 1. Cross-section view of the deformation fields for the monolithic and MG/Cu composites at different velocities. The amorphous and crystalline phases are colored according to atomic shear strain and CNA, respectively. The substrates were at an initial temperature of 100 K.

2.2. Penetration and Projectile Deformation

The vertical center-of-mass position of the projectile (Z_p) after impact was determined by calculating its z-coordinate, using the upper surface of the substrate as a reference. As shown in Figure 2a, higher impact velocities result in deeper penetration, with both substrates exhibiting similar behavior. The degree of horizontal dispersion of the projectile (Δd) was calculated as the difference between its post-impact radius and initial radius, with the results presented in Figure 2b. The projectile undergoes horizontal disintegration upon impact, consistent with previous studies on high-velocity impacts [21,26]. At lower velocities, Δd is comparable for both substrates. However, as the impact velocity increases, Δd becomes slightly higher for the monolithic MG substrate. This behavior can be attributed to the more homogeneous structure of the monolithic MG, which promotes a more uniform horizontal fragmentation of the projectile, as seen in the deformation fields in Figure 1.

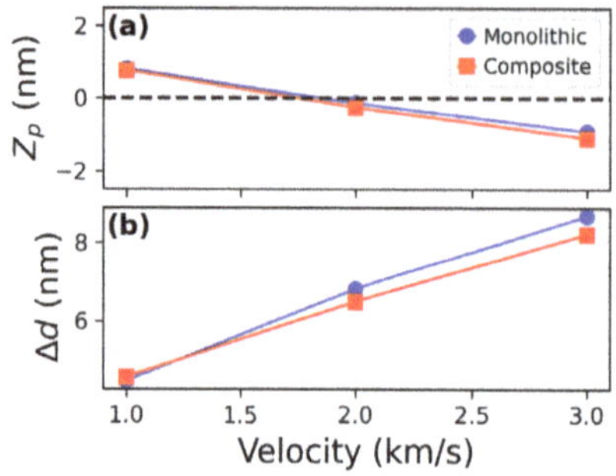

Figure 2. (**a**) Penetration depth (Z_p) into the substrate and (**b**) horizontal deformation of the projectile. The dotted line in Z_p represents the position of the substrate surface.

Further insight into the deformation of the projectile is provided in Figure 3, where the dotted line represents the upper surface of the substrate. It is evident that at lower velocities, the projectile largely retains its original shape, while higher velocities lead to significant deformation and deeper embedding into the substrate. Moreover, the final shape of the projectile is similar for both the monolithic and composite substrates, consistent with the trends observed in Z_p and Δd.

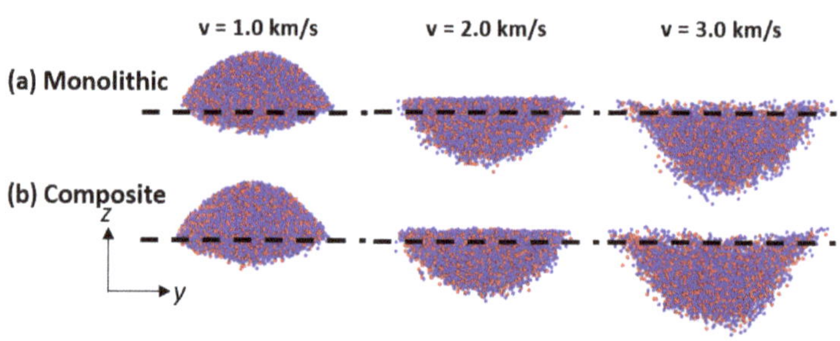

Figure 3. Projectile after 0.6 ns relaxation time for the (**a**) monolithic and (**b**) composite substrate. The dotted line represents the substrate upper surface. Red and blue colors represent Cu and Zr atoms, respectively.

High-velocity impacts result in substrates with elevated energy states. To investigate this effect, the variation in per-atom potential energy (ΔU) was quantified for both the monolithic and composite samples at different impact velocities. ΔU was calculated as the difference between the per-atom potential energy before and after impact. As shown in Figure 4a, ΔU increases with impact velocity due to the conversion of the projectile's kinetic energy into potential energy within the substrate. This increase in internal energy resembles the rejuvenation process commonly observed in MGs [46–48], although further mechanical characterization is required to confirm this analogy. The monolithic sample exhibits a greater energy variation compared to the composite, which can be attributed to the higher stiffness of the single-crystal Cu phase. This suggests that the crystalline phase in the MG/Cu composite substrate is less capable of undergoing plastic deformation due to its increased stiffness relative to the glass matrix. For reference, the Young's modulus of single-crystal Cu oriented along the ⟨100⟩ direction is approximately 100 GPa [49], whereas CuZr MGs typically range between 60–70 GPa, depending on atomic composition [36,50]. As a result, a greater amount of kinetic energy is required to produce a comparable increase in potential energy in the composite compared to the monolithic MG. Pile-up formation was also analyzed by measuring the maximum pile-up height relative to the substrate surface. As shown in Figure 4b, higher impact velocities lead to larger pile-ups. The monolithic sample exhibits lower pile-up heights than the composite, reflecting its reduced capacity for plastic deformation and atomic rearrangement in the bulk. This results in fewer atoms being displaced and accumulated at the surface. It is important to note that the pressure exerted by the projectile, along with the associated deformation, may modify the free volume and atomic structure of the amorphous phase through atomic rearrangements [51–53], which can enhance bulk deformation while reducing surface pile-up formation.

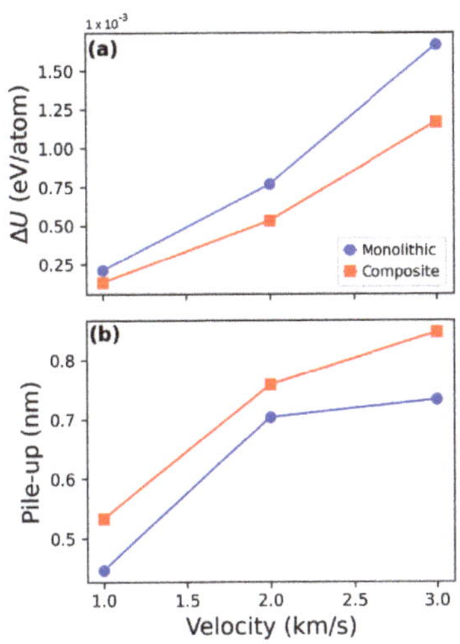

Figure 4. (**a**) Increase of the substrate per-atom potential energy and (**b**) crater pile-up after impact.

2.3. *Quantification of Plastic Deformation*

The high deformation fields induced significant plastic deformation in the samples. Plasticity in the amorphous phase was quantified using the degree of strain localization (ψ) and participation ratio (ϕ), while dislocation density (ρ_d) and the population of HCP atoms (P_{HCP}) were measured for the crystalline phase. These values are presented in Figure 5. As shown, all quantities exhibit an increasing trend with impact velocity, indicating greater energy transfer from the projectile to the substrate. A comparison between the substrates reveals that the composite undergoes higher strain localization and a greater number of participating atoms. This can be attributed to the increased stiffness of the crystalline phase in the composite, which induces more plastic activity in the amorphous matrix. Plastic characterization of the crystalline phase in the MG/Cu composite shows that both dislocation density and stacking fault formation increase with impact velocity, consistent with typical deformation mechanisms observed in FCC metals [43–45].

The distribution of plastic deformation was quantified in the radial direction from the impact region. To this end, the degree of strain localization and the population of HCP atoms were measured. The number of stacking faults was not explicitly quantified due to the difficulty of counting them directly. The resulting curves are shown in Figure 6. As observed, ψ increases with impact velocity and gradually decreases toward the bulk. When comparing the two substrates, ψ shows higher values in the composite, supporting our previous observation that the increased stiffness of the crystalline phase promotes deformation in the glass matrix. This is further confirmed by the radial distributions of HCP populations shown in Figure 6c. The populations are relatively small, consistent with the total populations observed in Figure 5c. The increasing trend of P_{HCP} in the radial direction suggests a larger number of stacking faults, although an exact count of these defects is not possible. It is important to note that the bins for calculating P_{HCP} become larger with increasing radial distance (r), which results in a higher fraction of HCP atoms being included in the calculation, thus explaining the observed trend.

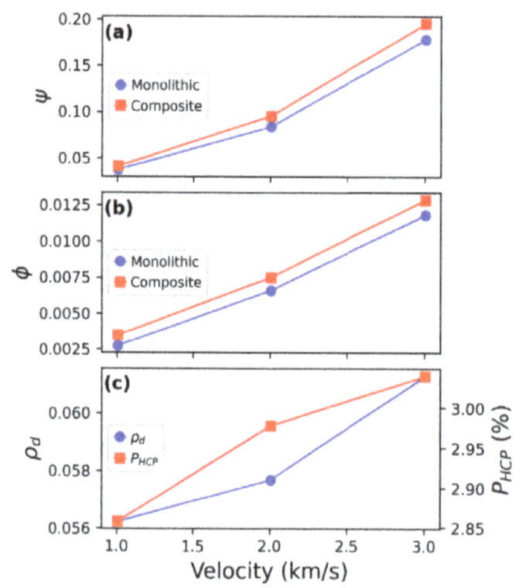

Figure 5. (**a**) Degree of strain localization (ψ) and (**b**) participation ratio (ϕ) for the entire substrate. (**c**) Dislocation density (ρ_d) and population of HCP atoms (P_{HCP}) for the crystalline phase.

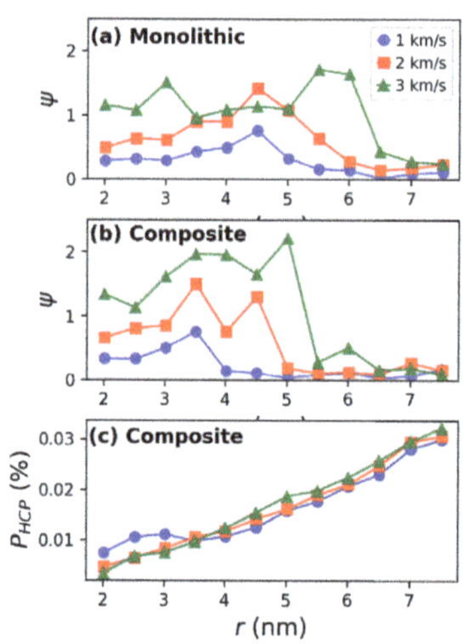

Figure 6. Radial distribution of the degree of strain localization (ψ) for the (**a**) monolithic and (**b**) composite substrate. (**c**) Radial distribution of the population of HCP atoms in the composite substrate.

2.4. Temperature Effect

To investigate the effect of initial temperature on the mechanical response of the substrate, five different initial temperatures were considered: 100 K, 200 K, 300 K, 400 K, and 500 K, with a fixed impact velocity of 2 km/s. The deformation fields for both substrates after impact are shown in Figure 7. A comparison across different initial temperatures reveals that higher kinetic energy leads to a greater number of atoms in the amorphous phase undergoing shear strain, consistent with previous studies [36,54]. However, the crystalline phase exhibits few differences. A comparison of both substrates at 500 K shows that the fraction of amorphous atoms experiencing high shear strain is larger in the composite substrate than in the monolithic one. This suggests that, despite the increased kinetic energy, the high stiffness of the crystalline phase inhibits its plastic deformation, leading to plastic events occurring predominantly in the glass matrix.

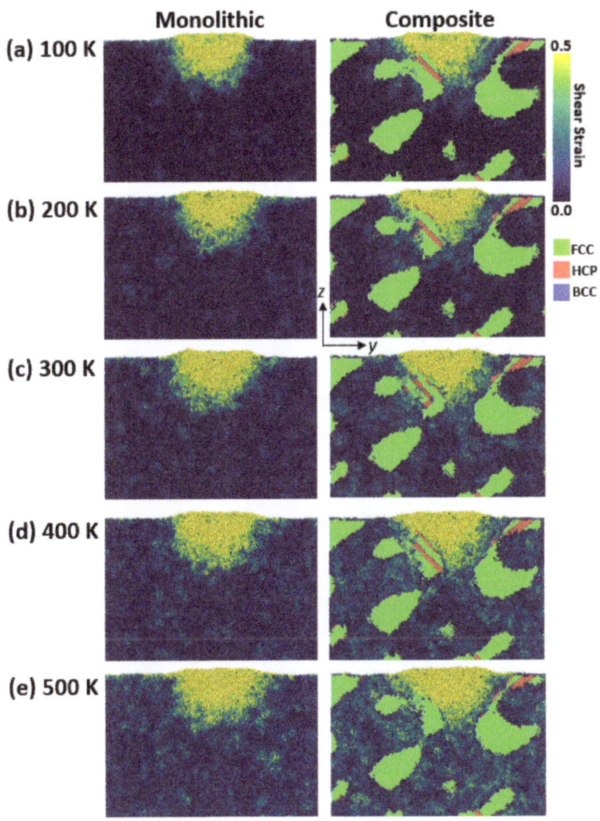

Figure 7. Cross-section view of the deformation fields for the monolithic and MG/Cu composites at different initial temperatures. The amorphous and crystalline phases are colored according to atomic shear strain and CNA, respectively.

The Z_p of the projectile was calculated for each case, with the results shown in Figure 8a. While Z_p is higher for the composite case, as discussed in the previous section, a slight decrease with temperature is observed. However, this trend is not statistically significant due to the low percentage difference (3%). Regarding the degree of horizontal fragmentation of the projectile after impact, an increasing trend is observed for both cases, indicating that higher kinetic energy enhances projectile disintegration in the lateral directions. However, there is almost no difference between the two substrates.

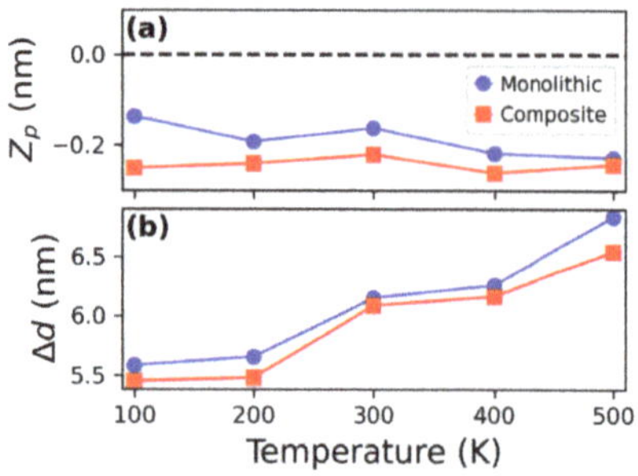

Figure 8. (**a**) Penetration depth (Z_p) into the substrate. (**b**) Horizontal deformation of the projectile when considering different substrate temperatures and an impact velocity of 2 km/s.

2.5. Relationship Between Plastic Deformation and Temperature

Plastic deformation was quantified for different initial temperatures, as shown in Figure 9. Both the degree of strain localization (ψ) and participation ratio (ϕ) exhibit increasing trends with temperature, whereas dislocation density and the population of HCP atoms show the opposite behavior. Interestingly, the participation ratio remains similar for both substrates at low temperatures but becomes significantly higher for the composite at elevated temperatures. This behavior reflects two key phenomena. First, the crystalline phase has a limited ability to undergo plastic deformation due to its higher stiffness, as previously discussed. Second, the glass matrix experiences enhanced atomic rearrangement, attributed to its disordered structure. Regarding plastic activity in the crystalline phase, both ρ_D and P_{HCP} decrease with temperature, reinforcing the observation that most plastic deformation occurs in the glass matrix rather than in the crystalline structure when compared with the results of ϕ. Moreover, previous studies have shown that higher temperatures lead to lower dislocation densities, as increased kinetic energy promotes dislocation emission, mobility, and absorption [35,55,56].

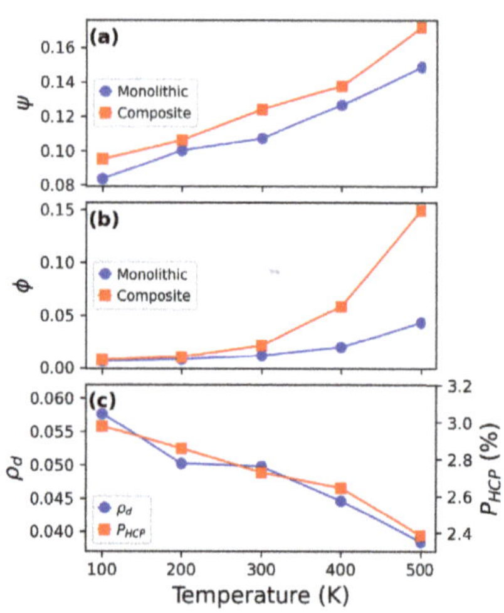

Figure 9. (**a**) Degree of strain localization (ψ) and (**b**) participation ratio (ϕ) for the entire substrate, and (**c**) dislocation density (ρ_d) and population of HCP atoms (P_{HCP}) for the crystalline phase for the substrate prepared at different initial temperatures. The considered impact velocity is 2 km/s.

2.6. Implications and Future Perspectives

The findings of this study offer valuable insights into the design of advanced protective materials for impact-prone environments, such as armor systems and aerospace structures exposed to space debris. The enhanced understanding of how MG and MG-based composites respond under extreme loading conditions can inform the development of lightweight, high-strength materials capable of dissipating large amounts of energy through localized plastic deformation. In particular, the superior plastic accommodation in MG matrices and the energy-dissipating capacity of composite architectures suggest promising avenues for tailoring impact-resistant coatings or layered shields. These results are especially relevant for spacecraft shielding, where minimizing mass while maximizing energy absorption is critical for mitigating damage from micrometeoroid and orbital debris impacts [57–59].

Despite the detailed atomic-level insights provided by MD simulations, the present study has some limitations. The simulations are restricted to nanometer-scale systems, which are difficult to explore experimentally. Moreover, the effects of interfacial properties and sample size were not fully explored and may influence deformation mechanisms in larger-scale systems. Future research could extend these findings by incorporating multi-scale modeling approaches and the evaluation of additional crystalline phase distributions. Additionally, understanding damage accumulation under repeated impacts or environmental effects such as oxidation or radiation would further enhance the applicability of MG-based composites in demanding operational environments.

3. Materials and Methods

MD simulations were performed using the LAMMPS (version 29Aug2024) package [60,61]. Cu-Zr atomic interactions were modeled using the interatomic potential developed by Borovikov et al. [62]. This potential accurately describes the amorphous phase of CuZr MGs and provides reliable values for the stacking fault energy of single Cu structures. An integration timestep of 1.0 fs was employed for all simulations.

The CuZrMG was prepared by constructing a crystalline box with dimensions of $51.8 \times 51.8 \times 5.18$ nm^3 and an atomic composition of 64% Cu and 36% Zr. The sample was equilibrated at 2500 K for 2 ns under zero pressure using periodic boundary conditions (PBCs). Subsequently, a cooling rate of 10^{11} K/s was applied to reduce the temperature to 100 K, followed by equilibration at zero pressure for 0.1 ns. The sample was replicated 10 times in the z-direction to obtain a larger structure. Finally, annealing at 600 K for 0.5 ns was performed to relieve stress at the artificial boundaries while maintaining zero pressure with PBCs.

A nanoporous MG structure was prepared using the algorithm proposed by Soyarslan et al. [63], based on the reaction-diffusion equation. Periodic waves $f(\vec{r})$ were randomly generated according to the following expression:

$$f(\vec{r}) = \sqrt{\frac{2}{N}\sum_{i=1}^{N} \cos(\vec{q}_i \cdot \vec{r} + \alpha_i)}. \tag{1}$$

Here, $\vec{q}_i$, $\vec{r}$, and α_i are the wave vector, the position vector, and a random phase, respectively. The wave vector is calculated as

$$\vec{q}_i = \frac{2\pi(h,k,l)}{L}, \tag{2}$$

where L is the simulation length, and h, k, and l are the Miller indices. Spinodal-like structures with translational periodicity are obtained according to the condition $H = \sqrt{h^2 + k^2 + l^2}$, where $|\vec{q}_i| = 2\pi H/L$ is a constant. Porosity is introduced when $f(\vec{r}) > \zeta$, where ζ is a free parameter. The resulting sample had a porosity of 30%. The porous regions were filled with single-crystalline Cu, with the [100], [010], and [001] orientations aligned along the x-, y-, and z-directions, respectively. The MG/Cu composite was annealed at 600 K for 0.5 ns under zero pressure with PBCs. Subsequently, the sample was equilibrated at five target temperatures, T_0: 100 K, 200 K, 300 K, 400 K, and 500 K. These MG/Cu composites were used as substrates for the impacts.

To simulate high-velocity impacts, a 5 nm-diameter MG nanoparticle was extracted from the bulk sample and positioned 5 nm above the substrate along the z-direction. Three different impact velocities were considered: 1 km/s, 2 km/s, and 3 km/s. A 1 nm-thick layer at the bottom of the substrate was set to zero force to prevent translational motion during the impact. Directly above this, a 1 nm-thick layer in the z-direction, as well as a 1 nm-thick layer along the outer edges of the x- and y-directions, was assigned a Langevin thermostat at a constant temperature T_0 to absorb the elastic waves generated during the impact. The inner region of the substrate was simulated under the NVE ensemble. A schematic of the projectile/substrate configuration is shown in Figure 10. Impacts were also performed on a monolithic MG for comparison purposes.

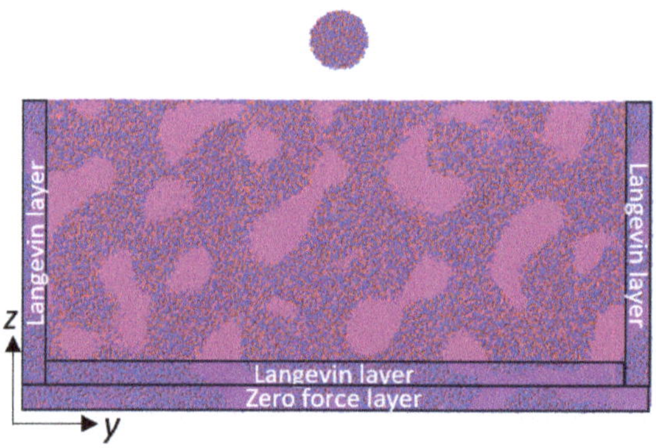

Figure 10. Scheme of the 1 nm-thick zero force and Langevin layers used in the simulations. The layers are not to scale for visualization purposes. Red and blue colors represent Cu and Zr atoms, respectively. Pink represents the crystalline phase.

Plasticity was quantified with the atomic shear strain [64]. The degree of strain localization was obtained as [65]

$$\psi = \sqrt{\frac{1}{N}\sum_{i=1}^{N}(\eta_i - \eta_{ave})^2}, \tag{3}$$

where N is the number of atoms, η_i is the von Mises strain of atom i obtained from the atomic shear strain, and η_{ave} is the average von Mises strain. The deformation participation ratio (ϕ) was obtained as [66,67]

$$\phi = \frac{N^{\eta>0.2}}{N_t}, \tag{4}$$

where $N^{\eta>0.2}$ is the number of atoms with von Mises strain greater than 0.2 and N_t is the total number of atoms. The dislocation density (ρ_d) was calculated using $\rho_d = L/V$, where L is the total dislocation length provided by the DXA algorithm (the sum of all the dislocation lengths) [68], and V is the total substrate volume. Crystalline structure was identified with the common neighbor analysis algorithm (CNA) [69], which was used to quantify the populations of HCP atoms. Visualization was performed using the OVITO (version 3.12.1) software [70].

4. Conclusions

The present study investigated the mechanical response of monolithic CuZr metallic glass (MG) and MG/Cu composite substrates under high-velocity impact using molecular dynamics simulations, considering variations in impact velocity and initial temperature.

Increasing the impact velocity led to deeper penetration and greater plastic deformation of both substrates. At higher velocities, the projectile exhibited more pronounced horizontal fragmentation in the monolithic substrate due to its homogeneous structure. The MG/Cu composite exhibited increased plasticity, attributed to the higher stiffness of the crystalline phase, which induced plastic deformation in the glass matrix. This difference in stiffness also influenced energy absorption, as the monolithic MG showed greater variations in potential energy due to its higher capacity for atomic rearrangement.

Temperature effects were primarily observed in the amorphous phase, where increasing the initial temperature enhanced shear strain localization and atomic mobility. In contrast, the crystalline phase exhibited lower variations in deformation behavior due to enhanced dislocation emission and absorption with temperature. This decrease in dislocation density and HCP atom population indicated that plastic activity shifted predominantly to the amorphous phase at elevated temperatures.

In summary, the MG/Cu composite displayed increased plasticity and lower energy variation, with the crystalline phase exhibiting little deformation. This indicates that the energy transferred from the projectile to the monolithic sample results in higher energy states rather than plastic deformation. Furthermore, higher temperatures induced more localized plasticity within the amorphous matrix, emphasizing the role of atomic mobility and structural disorder in accommodating deformation. These findings highlight the interplay between impact velocity, temperature, and material composition in governing plastic deformation mechanisms, providing insight into the response of MG-based composites under extreme loading conditions.

Supplementary Materials: The following supporting information can be downloaded at https://www.mdpi.com/article/10.3390/inorganics13050141/s1, the supplementary materials are LAMMPS scripts for the computational simulations.

Author Contributions: N.A.: conceptualization, software, investigation, methodology, visualization, formal analysis, writing—review and editing—original draft, resources. J.W.: formal analysis, visualization. P.L.-P.: investigation, visualization. All authors have read and agreed to the published version of the manuscript.

Funding: Project supported by the Competition for Research Regular Projects, year 2023, code LPR23-05, Universidad Tecnológica Metropolitana. Powered@NLHPC: This research was partially supported by the supercomputing infrastructure of the NLHPC (ECM-02).

Data Availability Statement: The original contributions presented in this study are included in the Supplementary Material. Further inquiries can be directed to the corresponding author.

Conflicts of Interest: The authors declare no conflicts of interest.

References

1. Ryan, S.; Schaefer, F.; Destefanis, R.; Lambert, M. A ballistic limit equation for hypervelocity impacts on composite honeycomb sandwich panel satellite structures. *Adv. Space Res.* **2008**, *41*, 1152–1166. [CrossRef]
2. Pieters, C.M.; Noble, S.K. Space weathering on airless bodies. *J. Geophys. Res. Planets* **2016**, *121*, 1865–1884. . [CrossRef]
3. Kılıc, N.; Bedir, S.; Erdik, A.; Ekici, B.; Tasdemirci, A.; Guden, M. Ballistic behavior of high hardness perforated armor plates against 7.62 mm armor piercing projectile. *Mater. Des.* **2014**, *63*, 427–438. [CrossRef]
4. Cour-Palais, B.G. Hypervelocity impact in metals, glass and composites. *Int. J. Impact Eng.* **1987**, *5*, 221–237. [CrossRef]
5. Safri, S.; Sultan, M.; Yidris, N.; Mustapha, F. Low velocity and high velocity impact test on composite materials–a review. *Int. J. Eng. Sci* **2014**, *3*, 50–60.
6. Collins, G.S.; Melosh, H.J.; Ivanov, B.A. Modeling damage and deformation in impact simulations. *Meteorit. Planet. Sci.* **2004**, *39*, 217–231. [CrossRef]
7. Seaman, L.; Curran, D.R.; Shockey, D.A. Computational models for ductile and brittle fracture. *J. Appl. Phys.* **1976**, *47*, 4814–4826. [CrossRef]
8. Wensink, H.; Elwenspoek, M.C. A closer look at the ductile-brittle transition in solid particle erosion. *Wear* **2002**, *253*, 1035–1043. [CrossRef]
9. Michel, Y.; Chevalier, J.M.; Durin, C.; Espinosa, C.; Malaise, F.; Barrau, J.J. Hypervelocity impacts on thin brittle targets: Experimental data and SPH simulations. *Int. J. Impact Eng.* **2006**, *33*, 441–451. [CrossRef]
10. Hao, G.; Dong, X.; Du, M.; Li, Z.; Dou, Z. A comparative study of ductile and brittle materials due to single angular particle impact. *Wear* **2019**, *428–429*, 258–271. [CrossRef]
11. Pham, T.M.; Hao, H. Influence of global stiffness and equivalent model on prediction of impact response of RC beams. *Int. J. Impact Eng.* **2018**, *113*, 88–97. [CrossRef]
12. Hassani, M.; Veysset, D.; Nelson, K.A.; Schuh, C.A. Material hardness at strain rates beyond 106 s-1 via high velocity microparticle impact indentation. *Scr. Mater.* **2020**, *177*, 198–202. [CrossRef]
13. Cai, J.; Griesbach, C.; Ahnen, S.G.; Thevamaran, R. Dynamic Hardness Evolution in Metals from Impact Induced Gradient Dislocation Density. *Acta Mater.* **2023**, *249*, 118807. [CrossRef]
14. Fortov, V.; Kim, V.; Lomonosov, I.; Matveichev, A.; Ostrik, A. Numerical modeling of hypervelocity impacts. *Int. J. Impact Eng.* **2006**, *33*, 244–253. [CrossRef]
15. Tan, V.; Ching, T. Computational simulation of fabric armour subjected to ballistic impacts. *Int. J. Impact Eng.* **2006**, *32*, 1737–1751. [CrossRef]
16. Kilic, N.; Ekici, B.; Hartomacıoglu, S. Determination of penetration depth at high velocity impact using finite element method and artificial neural network tools. *Def. Technol.* **2015**, *11*, 110–122. [CrossRef]
17. Žmindák, M.; Pelagić, Z.; Pastorek, P.; Močilan, M.; Vybošťok, M. Finite Element Modelling of High Velocity Impact on Plate Structures. *Procedia Eng.* **2016**, *136*, 162–168. [CrossRef]
18. Yehia Abdel-Nasser, A.M.E.; Al-Mallah, I. Impact analysis of composite laminate using finite element method. *Ships Offshore Struct.* **2017**, *12*, 219–226. [CrossRef]
19. Madhukar, A.; Ostoja-Starzewski, M. Finite Element Methods in Human Head Impact Simulations: A Review. *Ann. Biomed. Eng.* **2019**, *47*, 1832–1854. [CrossRef]
20. Song, G.; Yang, H.; Hogan, C.J. Thermal energy evolution and mechanical deformation of monocrystalline yttria-stabilized zirconia nanoparticles in aerosol deposition processes. *Appl. Surf. Sci.* **2022**, *585*, 152603. [CrossRef]
21. Burlison, S.; Becker, M.F.; Kovar, D. A molecular dynamics study of the effects of velocity and diameter on the impact behavior of zinc oxide nanoparticles. *Model. Simul. Mater. Sci. Eng.* **2023**, *31*, 075008. [CrossRef]
22. Song, G.; Hogan, C.J. Crystal grain size effects and crystallinity dynamics during supersonic particle impacts. *Int. J. Plast.* **2023**, *170*, 103758. [CrossRef]
23. Amigo, N. Deposition of droplets on metallic glass substrates: Thermophysical and structural characterization. *J. Mol. Liq.* **2025**, *421*, 126879. [CrossRef]
24. Chitrakar, T.V.; Becker, M.F.; Kovar, D. Influence of Crystallographic Orientation on the Deformation of Ag Nanoparticles During High-Speed Impact. *J. Therm. Spray Technol.* **2023**, *32*, 2683–2700. [CrossRef]
25. Dutta, A. Surface damage of CuZr metallic glass by hypervelocity nano-projectile: A molecular dynamics study. *Comput. Mater. Sci.* **2018**, *141*, 41–48. [CrossRef]
26. Wachter, J.; Amigo, N.; Gutiérrez, G.; Zúñiga, A. Atomic scale study of the impact of metallic glass nanoparticles at high velocities. *J. Non-Cryst. Solids* **2024**, *628*, 122866. [CrossRef]
27. Dwivedi, P.; Fraile, A.; Polcar, T. Tungsten wall cratering under high-velocity dust impacts: Influence of impact angle and temperature. *J. Nucl. Mater.* **2024**, *600*, 155289. [CrossRef]

28. Greer, A.L. Metallic glasses…on the threshold. *Mater. Today* **2009**, *12*, 14–22. [CrossRef]
29. Schroers, J. Processing of Bulk Metallic Glass. *Adv. Mater.* **2010**, *22*, 1566–1597. [CrossRef]
30. Chen, M. A brief overview of bulk metallic glasses. *NPG Asia Mater.* **2011**, *3*, 82–90. [CrossRef]
31. Wu, F.F.; Chan, K.C.; Jiang, S.S.; Chen, S.H.; Wang, G. Bulk metallic glass composite with good tensile ductility, high strength and large elastic strain limit. *Sci. Rep.* **2014**, *4*, 5302. [CrossRef]
32. Qiao, J.; Jia, H.; Liaw, P.K. Metallic glass matrix composites. *Mater. Sci. Eng. Rep.* **2016**, *100*, 1–69. [CrossRef]
33. Sha, Z.D.; Branicio, P.S.; Lee, H.P.; Tay, T.E. Strong and ductile nanolaminate composites combining metallic glasses and nanoglasses. *Int. J. Plast.* **2017**, *90*, 231–241. [CrossRef]
34. Fu, X.; Wang, G.; Wu, Y.; Song, W.; Shek, C.H.; Zhang, Y.; Shen, J.; Ritchie, R.O. Compressive ductility and fracture resistance in CuZr-based shape-memory metallic-glass composites. *Int. J. Plast.* **2020**, *128*, 102687. [CrossRef]
35. Yaghoobi, M.; Voyiadjis, G.Z. The effects of temperature and strain rate in fcc and bcc metals during extreme deformation rates. *Acta Mater.* **2018**, *151*, 1–10. [CrossRef]
36. Amigo, N.; Careglio, C.A.; Ardiani, F.; Manelli, A.; Tramontina, D.R.; Bringa, E.M. Thermal effects on the mechanical behavior of CuZr metallic glasses. *Appl. Phys. A* **2024**, *130*, 616. [CrossRef]
37. Amigo, N. Mechanical characterization of nanocrystalline Cu-Ag alloys subjected to shear deformation. *Mol. Simul.* **2024**, *50*, 1411–1419. [CrossRef]
38. Reddy, K.V.; Deng, C.; Pal, S. Dynamic characterization of shock response in crystalline-metallic glass nanolaminates. *Acta Mater.* **2019**, *164*, 347–361. [CrossRef]
39. Zhang, J.; Wang, X.; Zhang, M.; Deng, L.; Gong, P. The shock forming process of Cu50Zr50 metallic glasses studied via molecular dynamics simulation. *J. Non-Cryst. Solids* **2022**, *584*, 121518. [CrossRef]
40. Song, W.; Yu, Y.; Guan, Y. Role of void shape on shock responses of nanoporous metallic glasses via molecular dynamics simulation. *Int. J. Mech. Sci.* **2022**, *218*, 107076. [CrossRef]
41. Song, W.; Meng, Y.; Xiao, L.; Liu, S. Tuning dynamic mechanical properties of Cu50Zr50 nanoglasses/ nanopolycrystalline Cu composites investigated by molecular dynamics simulation. *J. Non-Cryst. Solids* **2024**, *625*, 122715. [CrossRef]
42. Jami, H.; Jabbarzadeh, A. Molecular simulation of high-velocity deposition of titanium dioxide nanoparticles on titanium. *Appl. Surf. Sci.* **2021**, *542*, 148567. [CrossRef]
43. Stukowski, A.; Albe, K.; Farkas, D. Nanotwinned fcc metals: Strengthening versus softening mechanisms. *Phys. Rev. B* **2010**, *82*, 224103. [CrossRef]
44. Asadi, E.; Zaeem, M.A.; Moitra, A.; Tschopp, M.A. Effect of vacancy defects on generalized stacking fault energy of fcc metals. *J. Physics: Condens. Matter* **2014**, *26*, 115404. [CrossRef] [PubMed]
45. Amigo, N. Mechanical Properties of Copper with Dendritic Silver Inclusions: Insights from Molecular Dynamics Simulations. *Comput. Mater. Contin.* **2024**, *81*, 3665–3678. [CrossRef]
46. Priezjev, N.V. Aging and rejuvenation during elastostatic loading of amorphous alloys: A molecular dynamics simulation study. *Comput. Mater. Sci.* **2019**, *168*, 125–130. [CrossRef]
47. Wang, M.; Liu, H.; Li, J.; Jiang, Q.; Yang, W.; Tang, C. Thermal-pressure treatment for tuning the atomic structure of metallic glass Cu-Zr. *J. Non-Cryst. Solids* **2020**, *535*, 119963. [CrossRef]
48. Amigo, N. Cryogenic thermal cycling rejuvenation in metallic glasses: Structural and mechanical assessment. *J. Non-Cryst. Solids* **2022**, *596*, 121850. [CrossRef]
49. Gao, Y.; Wang, H.; Zhao, J.; Sun, C.; Wang, F. Anisotropic and temperature effects on mechanical properties of copper nanowires under tensile loading. *Comput. Mater. Sci.* **2011**, *50*, 3032–3037. [CrossRef]
50. Amigo, N.; Valencia, F. Mechanical and structural assessment of CuZr metallic glasses rejuvenated by thermal-pressure treatments. *Comput. Mater. Sci.* **2021**, *198*, 110681. [CrossRef]
51. Heggen, M.; Spaepen, F.; Feuerbacher, M. Creation and annihilation of free volume during homogeneous flow of a metallic glass. *J. Appl. Phys.* **2004**, *97*, 033506. [CrossRef]
52. Yoo, B.G.; Kim, J.Y.; Kim, Y.J.; Choi, I.C.; Shim, S.; Tsui, T.Y.; Bei, H.; Ramamurty, U.; Jang, J.I. Increased time-dependent room temperature plasticity in metallic glass nanopillars and its size-dependency. *Int. J. Plast.* **2012**, *37*, 108–118. [CrossRef]
53. Qiu, Y.; Xu, C.; Fu, E.; Wang, P.; Du, J.; Hu, Z.; Yan, X.; Cao, X.; Wang, Y.; Shao, L. Mechanisms for the free volume tuning the mechanical properties of metallic glass through ion irradiation. *Intermetallics* **2018**, *101*, 173–178. [CrossRef]
54. Jafary-Zadeh, M.; Tavakoli, R.; Srolovitz, D.J.; Zhang, Y.W. Thermally induced failure mechanism transition and its correlation with short-range order evolution in metallic glasses. *Extrem. Mech. Lett.* **2016**, *9*, 215–225. [CrossRef]
55. Rohith, P.; Sainath, G.; Srinivasan, V. Effect of size, temperature and strain rate on dislocation density and deformation mechanisms in Cu nanowires. *Phys. Condens. Matter* **2019**, *561*, 136–140. [CrossRef]

56. Saffarini, M.H.; Voyiadjis, G.Z.; Ruestes, C.J. Temperature effect on nanoporous gold under uniaxial tension and compression. *Comput. Mater. Sci.* **2021**, *200*, 110766. [CrossRef]
57. Grossman, E.; Gouzman, I.; Verker, R. Debris/Micrometeoroid Impacts and Synergistic Effects on Spacecraft Materials. *MRS Bull.* **2010**, *35*, 41–47. [CrossRef]
58. Schonberg, W.P. Protecting Earth-orbiting spacecraft against micro-meteoroid/orbital debris impact damage using composite structural systems and materials: An overview. *Adv. Space Res.* **2010**, *45*, 709–720. [CrossRef]
59. Rakib, M.A.; Smith, S.T.; Tafsirojjaman, T. A review of shielding systems for protecting off-earth structures from micrometeoroid and orbital debris impact. *Acta Astronaut.* **2024**, *223*, 404–425. [CrossRef]
60. Plimpton, S. Fast Parallel Algorithms for Short-Range Molecular Dynamics. *J. Comput. Phys.* **1995**, *117*, 1–19. [CrossRef]
61. Thompson, A.P.; Aktulga, H.M.; Berger, R.; Bolintineanu, D.S.; Brown, W.M.; Crozier, P.S.; In 't Veld, P.J.; Kohlmeyer, A.; Moore, S.G.; Nguyen, T.D.; et al. LAMMPS–A flexible simulation tool for particle-based materials modeling at the atomic, meso, and continuum scales. *Comp. Phys. Comm.* **2022**, *271*, 108171. [CrossRef]
62. Borovikov, V.; Mendelev, M.I.; King, A.H. Effects of stable and unstable stacking fault energy on dislocation nucleation in nano-crystalline metals. *Model. Simul. Mater. Sci. Eng.* **2016**, *24*, 085017. [CrossRef]
63. Soyarslan, C.; Bargmann, S.; Pradas, M.; Weissmüller, J. 3D stochastic bicontinuous microstructures: Generation, topology and elasticity. *Acta Mater.* **2018**, *149*, 326–340. [CrossRef]
64. Shimizu, F.; Ogata, S.; Li, J. Theory of Shear Banding in Metallic Glasses and Molecular Dynamics Calculations. *Mater. Trans.* **2007**, *48*, 2923–2927. [CrossRef]
65. Cheng, Y.Q.; Cao, A.J.; Ma, E. Correlation between the elastic modulus and the intrinsic plastic behavior of metallic glasses: The roles of atomic configuration and alloy composition. *Acta Mater.* **2009**, *57*, 3253–3267. [CrossRef]
66. Shi, Y.; Falk, M.L. Strain Localization and Percolation of Stable Structure in Amorphous Solids. *Phys. Rev. Lett.* **2005**, *95*, 095502. [CrossRef]
67. Adibi, S.; Branicio, P.S.; Joshi, S.P. Suppression of Shear Banding and Transition to Necking and Homogeneous Flow in Nanoglass Nanopillars. *Sci. Rep.* **2015**, *5*, 15611. [CrossRef]
68. Stukowski, A. Structure identification methods for atomistic simulations of crystalline materials. *Model. Simul. Mater. Sci. Eng.* **2012**, *20*, 045021. [CrossRef]
69. Faken, D.; Jónsson, H. Systematic analysis of local atomic structure combined with 3D computer graphics. *Comput. Mater. Sci.* **1994**, *2*, 279–286. [CrossRef]
70. Stukowski, A. Visualization and analysis of atomistic simulation data with OVITO–the Open Visualization Tool. *Model. Simul. Mater. Sci. Eng.* **2010**, *18*, 015012. [CrossRef]

inorganics

MDPI

Article

Uncooled Microbolometers Based on Nitrogen-Doped Hydrogenated Amorphous Silicon-Germanium (a-SiGe:H,N)

Oscar Velandia [1], Alfonso Torres [1], Alfredo Morales [1], Luis Hernández [1], Alberto Luna [2], Karim Monfil [2], Javier Flores [3,4], Gustavo M. Minquiz [3,4], Ricardo Jiménez [1] and Mario Moreno [1,*]

[1] Electronics Coordination, Instituto Nacional de Astrofísica, Óptica y Electrónica, Tonantzintla, Puebla 72840, Mexico; oscarveca@gmail.com (O.V.); atorres@inaoep.mx (A.T.); alfredom@inaoep.mx (A.M.); luish@inaoep.mx (L.H.)

[2] Centro de Investigaciones en Dispositivos Semiconductores (CIDS-ICUAP), Benemérita Universidad Autónoma de Puebla (BUAP), Col. San Manuel, Cd. Universitaria, Av. San Claudio y 14 Sur, Edificios IC5 y IC6, Puebla 72570, Mexico; jose.luna@correo.buap.mx (A.L.); karim.monfil@correo.buap.mx (K.M.)

[3] Ingeniería Mecánica, Tecnológico Nacional de México/I.T. Puebla, Av. Tecnológico #420 Col. Maravillas, Puebla 72220, Mexico; javier.flores@puebla.tecnm.mx (J.F.); gminquiz@yahoo.com (G.M.M.)

[4] Área de Ingeniería, Benemérita Universidad Autónoma de Puebla, Ciudad Universitaria, Blvd. Valsequillo y Esquina Av. San Claudio s/n, Col. San Manuel, Puebla 72570, Mexico

* Correspondence: mmoreno@inaoep.mx; Tel.: +52-(222)-266-31-00

Abstract: An uncooled microbolometer is a thermal sensor consisting of a membrane suspended from the substrate to provide thermal insulation. Typically, the membrane is composed of a stack of three films integrated by a supporting film, an IR sensing film, and an IR absorbing film. However, the above increases the thickness of the device and affects its mechanical stability and thermal mass, thereby reducing its performance. One solution is to use a single film as a membrane with both IR sensing and IR absorbing properties. In this regard, this work presents the fabrication and evaluation of uncooled microbolometers using nitrogen-doped hydrogenated amorphous silicon-germanium (a-SiGe:H,N) as a single IR-absorber/IR sensing membrane. The films were deposited via low frequency Plasma Enhanced Chemical Vapor Deposition (PECVD) at 200 °C. Three microbolometer configurations were fabricated using a-SiGe:H,N films deposited from a SiH_4, GeH_4, N_2, and H_2 gas mixture with different SiH_4 and GeH_4 flow rates and, consequently, with different properties, such as temperature coefficient of resistance (*TCR*) and conductivity at room temperature. The microbolometer that exhibited the best performance achieved a voltage responsivity of 7.26×10^5 V/W and a *NETD* of 22.35 mK at 140 Hz, which is comparable to state-of-the-art uncooled infrared (IR) sensors. These results confirm that the optimization of the deposition parameters of the a-SiGe:H,N films significantly affects the microbolometers final performance, enabling an optimal balance between thermal sensitivity (*TCR*) and conductivity.

Keywords: PECVD; microbolometer; infrared; amorphous; silicon; germanium; TCR; conductivity; responsivity; detectivity; NEP; NETD

1. Introduction

Uncooled microbolometers have proven to be a crucial technology for thermal detection systems due to their ability to detect infrared radiation without the necessity of cryogenic cooling systems, reducing costs and facilitating their integration into compact and portable devices. These sensors are essential in applications such as surveillance [1], medical diagnostics [2], industrial inspection [3], automotive systems [4], firefighting [5], and agriculture [6]. Their importance continues growing thanks to their capability to operate under adverse conditions and provide efficient portable solutions [7].

An uncooled microbolometer is a thermal sensor used as a pixel in large infrared (IR) focal plane arrays (IRFPAs). This device is a temperature-dependent resistor, where the absorption of IR radiation causes an increment in its temperature, and consequently it experiences a change in its electrical resistance. The temperature coefficient of resistance (*TCR*) indicates the percentage of the microbolometer electrical resistance that changes per degree Kelvin in temperature (%/K).

An uncooled microbolometer consists of a suspended membrane from the substrate to provide thermal isolation. Basically, the membrane is composed of a film stack integrated by a supporting film, typically of silicon nitride (SiN_x) [8], an IR sensing film, typically of hydrogenated amorphous silicon (a-Si:H) [9], and an IR absorbing film, also of SiN_x, which is an excellent IR absorber [10]. However, the above increases the steps in the sensor fabrication process, increases the device thickness, and affects its mechanical stability and thermal mass, thereby reducing its performance. One solution is the use of just one film as a membrane, with both properties of IR sensing and IR absorbing.

Vanadium oxide (VOx) was the first IR sensing element used in uncooled microbolometers [11] with a *TCR* in the range of −2.0%/K to −2.7%/K, and more recently intrinsic a-Si:H has been adopted due to its very large *TCR* values (up to −9%/K). However, a-Si:H also has a very large electrical resistance, which implies incompatibility with the input impedance of the readout integrated circuits (ROICs) [8]. Therefore, boron has been employed to dope a-Si:H (a-Si:H,B) to reduce its electrical resistance, but also a reduction in *TCR* to −3%/K is obtained [12].

In our previous work we have studied intrinsic hydrogenated amorphous silicon-germanium (a-SiGe:H) alloys, which present high values of *TCR* (above −4%/K) and moderate electrical resistance, depending on the silicon and germanium content in the alloy [13]. In this context, nitrogen-doped hydrogenated amorphous silicon-germanium (a-SiGe:H,N) is a promising material for being used as an IR sensing/IR absorbing membrane due to its high IR absorption, high *TCR*, good electrical performance, and compatibility with silicon Complementary Metal-Oxide-Semiconductor (CMOS) transistor technology. The above enables the production of simpler, low-cost, and high-performance thermal sensors [14].

The Plasma Enhanced Chemical Vapor Deposition (PECVD) technique is essential for depositing a-SiGe:H,N films, as it allows precise control over the material properties by adjusting the deposition conditions, such as gas flow rates, RF power, chamber pressure, and substrate temperature. These adjustments directly impact the device *TCR*, responsivity, and conductivity, which are essential factors for improving thermal sensitivity and reducing noise levels. Additionally, the ability to fine-tune the material structure through PECVD enables the films to absorb efficiently in the LWIR range (8–14 μm), making them ideal for thermal imaging applications [15,16]. This study aims to optimize the properties of the microbolometer by evaluating three configurations of a-SiGe:H,N films, selected based on their optical and electrical performance. The results provide clear guidelines for developing advanced uncooled infrared sensors compatible with the silicon CMOS technology and facilitating their integration into advanced industrial IRFPAs.

2. Results and Discussion

2.1. Characterization of the Thermo-Sensing Films

In this section, the main characteristics of the thermos-sensing films are presented. Figure 1 shows the absorption coefficient spectra of the a-SiGe:H,N films, obtained by Fourier transform infrared (FTIR) spectroscopy. For this characterization, the a-SiGe:H,N thin films were deposited on 1 square inch crystalline silicon (c-Si) substrates, specifically designed for optical characterization, using a FTIR spectrometer Nicolet iS50 (Thermo Scientific, Waltham, MA, USA). The thicknesses of the a-SiGe:H,N films studied in this work are the following: film #1 has a thickness of 327 nm, film #2 has a thickness of 361 nm, and film #3 has a thickness of 392 nm, according to measurements performed with a mechanical profiler.

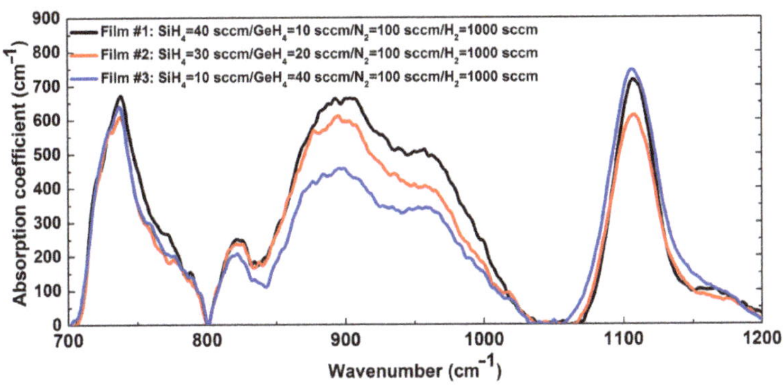

Figure 1. Absorption coefficient spectra of the a-SiGe:H,N films deposited with different flow rates of SiH_4 and GeH_4, maintaining constant the N_2 and H_2 flow rates.

It can be observed in the spectra in the range of 800 cm^{-1} to 1000 cm^{-1} (corresponding to a wavelength of 10 μm to 12.5 μm) that film #1 has the largest absorption coefficient, while film #3 has the lowest. That region is of interest because the microbolometers are designed to detect radiation at 10 μm, which is the wavelength at which the human body emits IR radiation. Therefore, we would expect the microbolometer to present a better performance with film #1.

Atomic force microscopy (AFM) was used to characterize the surface average roughness of the three a-SiGe:H,N films. Figure 2 shows 3-D images of the surface of the films measured in areas of 4 μm × 4 μm employing an AFM (EasyScan, Nanosurf, Liestal, Switzerland) where it can be observed that film #1 has a sharper surface, while film #3 has the roughest one. Table 1 shows the average roughness for the three films, where effectively it is shown that film #1 has the lower average roughness. This is of importance for the fabrication of microbolometers, where a sharp thermos-sensing film is required to have suitable electrical contact with the metal electrodes.

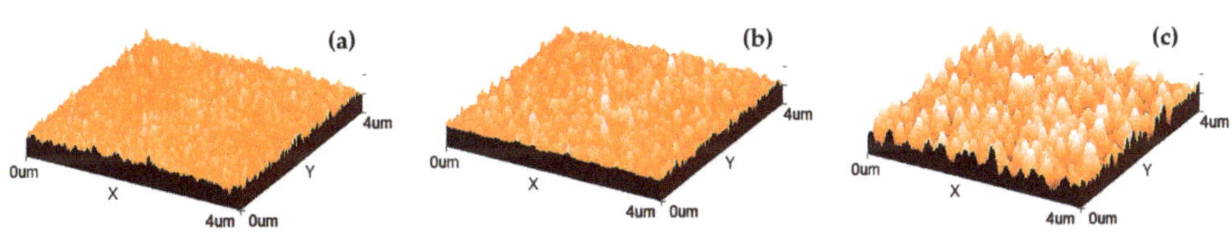

Figure 2. Atomic force microscopy (AFM) images of the a-SiGe:H,N films: (**a**) Film #1, (**b**) Film #2, (**c**) Film #3.

Table 1. Average roughness of the three a-SiGe:H,N thermo-sensing films measured with AFM (Atomic Force Microscopy). The voltage responsivity, current responsivity, and room temperature resistance of the microbolometers are also shown with the three different films.

Microbolometer Films	Gases Flow Rates Used for Deposition (sccm)	Average Roughness (nm)	Voltage Responsivity $\Re_V$ (V/W)	Current Responsivity $\Re_I$ (A/W)	Room Temperature Resistance (Ω)
Film #1	SiH_4 = 40 sccm/GeH_4 = 10 sccm/ N_2 = 100 sccm/H_2 = 1000 sccm	4.51	7.26×10^5	5.25×10^2	2.76×10^3
Film #2	SiH_4 = 30 sccm/GeH_4 = 20 sccm/ N_2 = 100 sccm/H_2 = 1000 sccm	5.68	4.99×10^5	3.99×10^{-1}	1.57×10^8
Film #3	SiH_4 = 10 sccm/GeH_4 = 40 sccm/ N_2 = 100 sccm/H_2 = 1000 sccm	8.40	8.16×10^5	2.60×10^{-1}	5.72×10^6

2.2. Characterization of Micro-Bolometers

In this section, the results obtained during the characterization of the microbolometer are presented. Key parameters such as responsivity of both voltage and current ($\Re_{V,\,i}$), power spectral density (PSD), noise equivalent power (NEP), normalized detectivity (D*), and noise equivalent temperature difference (NETD) were evaluated. These parameters are essential for determining the device sensitivity and efficiency, which are crucial for its performance in infrared detection applications. Tests were conducted under controlled conditions of temperature, vacuum, and radiation power, using specialized equipment to ensure accuracy in the results, which are detailed in the following sections. The data obtained provides valuable insights for optimizing future designs and enhancing device performance in various infrared detection applications.

2.3. Responsivity ($\Re_V$, $\Re_i$)

Responsivity indicates the microbolometers ability to convert infrared radiation into an electrical signal. It is calculated both in terms of voltage ($\Re_V$) and current ($\Re_i$) using the following equations:

$$\Re_V = \frac{\Delta V_{out}}{P_{IR}}\ (V/W) \tag{1}$$

$$\Re_i = \frac{\Delta I_{out}}{P_{IR}}\ (A/W) \tag{2}$$

where V_{out} and I_{out} represent the measured output voltage and current, respectively, and P_{IR} is the incident infrared radiation power [17,18]. Responsivity is a critical indicator in the characterization of thermal sensors, as it defines the efficiency with which the device can detect and convert radiation [10].

To measure these parameters, the microbolometer was placed in a cryostat (MMR Technologies, San Jose, CA, USA) in a vacuum environment of 40 mTorr that allows the characterization at room temperature conditions and IR illumination for current-voltage I(V) characterization, connected via microprobes. The I(V) measurements were conducted at 300 K, controlled and monitored with a Model 331 cryogenic temperature controller (Lake Shore, Westerville, OH, USA).

Temperature stabilization is necessary for the responsivity characterization to ensure that the change in electrical resistance is due to the IR radiation absorption and not to variations in the sensor temperature due to the environment. Also, this characterization is performed in a vacuum to eliminate undesired heat transference mechanisms, mainly to reduce heat transfer by convection.

The I(V) curve was obtained both in the dark and under IR illumination to evaluate the devices performance. Electrical characterization was performed using a 6517A electrometer (Keithley, Solon, OH, USA), controlled by a computer through LabView software. The voltage was applied from 0 to 1.5 V in the dark and under IR radiation. A higher excitation voltage can increase the responsivity; however, it was limited to 1.5 V to avoid damaging the sensors, such as a membrane breakdown.

The infrared radiation source was provided by a Kanthal Globar silicon carbide bar, whose radiation passes through a zinc selenide (ZnSe) window with 70% transmittance in the range of 0.6 to 20 µm. This radiation is further filtered by a 260 µm thick silicon wafer, restricting the spectrum to a region of approximately 1–15 µm. The incident radiation, measuring 220.5 nW, was directed toward the microbolometer using a mirror placed 20 cm away. The IR radiation was measured with an Oriel 71968 thermopile (Newport Corporation, Irvine, CA, USA), Figure 3 shows the experimental setup [19].

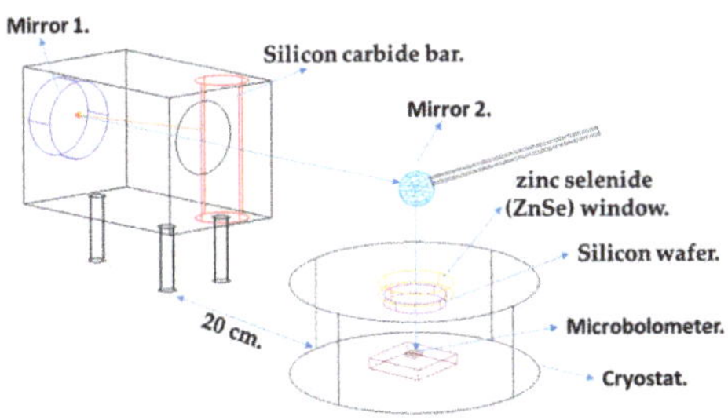

Figure 3. Experimental setup for microbolometer characterization under infrared radiation.

During the experiments, IR radiation was modulated using an optical chopper (Stanford Research Systems, Sunnyvale, CA, USA). Specifically for the responsivity measurements, the chopper frequency was fixed to 5 Hz to ensure maximum microbolometer response.

Also, to characterize the thermal response time of the microbolometer, the cutoff frequency was analyzed, defined as the point at which the output signal decreases by −3 dB from its maximum value. This 3 dB attenuation corresponds to a reduction to 70.7% of the maximum detected signal amplitude [20]. The cutoff frequency was found at 70 Hz, corresponding to a thermal response time of approximately 14 milliseconds.

Figure 4 shows the current-voltage curves obtained from three microbolometers with different thermo-sensing films deposited under different gas flow conditions, as was previously discussed. To evaluate the materials IR response, measurements were performed under two main conditions: in the dark and under IR illumination. In graphs (a), (b), and (c) of Figure 4, the variation in current as a function of the applied voltage is observed, comparing the microbolometers response under both dark and IR conditions.

The difference in current (ΔI) and voltage (ΔV) between dark and IR illumination allows the calculation of the device responsivity, expressed in A/W and V/W, which is an indicator of the device efficiency in converting incident radiation into an electrical signal. In addition to responsivity, relevant parameters such as the devices room temperature resistance are presented. Responsivity varies depending on the thermo-sensing films deposited with different SiH_4/GeH_4 gas flow rates, providing insights into how these factors affect the sensor sensitivity.

Table 1 shows the electrical parameters of three microbolometers with different a-SiGe:H,N thermo-sensing films deposited with different gas flow rates (see Table 1), which affects their responsivity and room temperature resistance. Voltage responsivity (V/W) and current responsivity (A/W) indicate the sensors ability to convert IR radiation into electrical signals of voltage and current, respectively, with higher values being indicative of greater sensitivity. The room temperature electrical resistance (Ω), measured for each type of microbolometer, provides information on the sensor compatibility with the input impedance of the silicon CMOS ROIC [21].

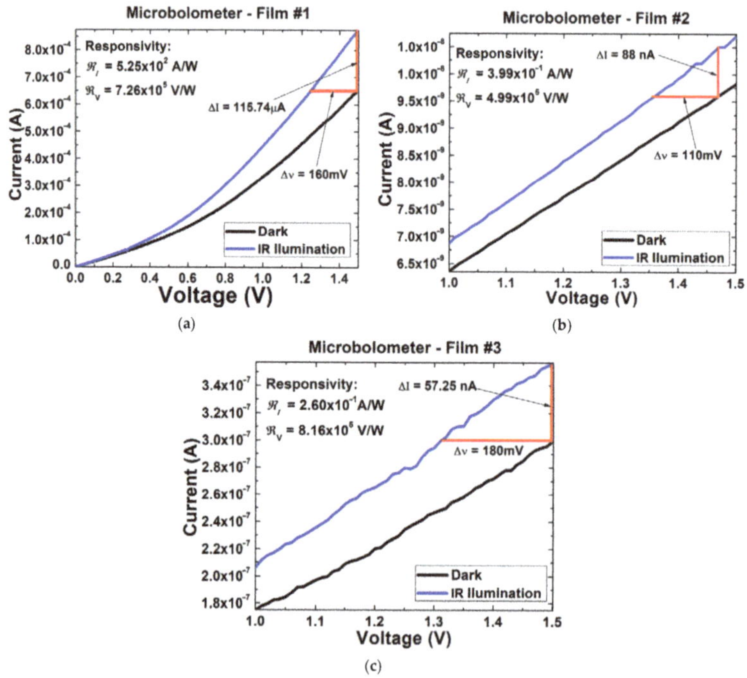

Figure 4. Current-voltage curves of three microbolometers with different a-SiGe:H,N thermo-sensing films under dark conditions and IR illumination: (**a**) microbolometer with film #1, (**b**) microbolometer with film #2, and (**c**) microbolometer with film #3.

The microbolometer with film #3 had a voltage responsivity of 8.16×10^5 V/W and a current responsivity of 2.6×10^{-1} A/W, along with a resistance of 5.72×10^6 Ω. This configuration shows the highest voltage responsivity but also the highest electrical resistance. The above can be related to the very high roughness of the thermo-sensing film and an unsuitable metal-semiconductor contact with the metal electrode.

On the other hand, the microbolometer with film #1 shows the highest current responsivity of 5.25×10^2 A/W and also high voltage responsivity of 7.26×10^5 V/W, but with the lowest resistance of 2.76×10^3. The low microbolometer resistance also contributes to a good balance between sensitivity and compatibility with the input impedance of the silicon CMOS ROIC. These data are essential for understanding how variations in the gas flow rates for the a-SiGe:H,N film deposition affect the microbolometer responsivity and electrical resistance properties.

2.4. Power Spectral Density (PSD)

Measurements were conducted using an SR780 spectrum analyzer, manufactured by Stanford Research Systems (SRS), under controlled conditions in a vacuum cryostat (40 mTorr) at a temperature of 300 K. This device is widely used for signal analysis in the frequency domain, allowing for precise power spectral density (*PSD*) measurements in units of $V_{rms}/\sqrt{Hz}$.

The equation used to describe the noise voltage of an open-circuit resistor is:

$$Vn = \sqrt{4kTR\Delta f} \left({}^{v_{rms}} / {}_{\sqrt{Hz}} \right) \quad (3)$$

where k is the Boltzmann constant (1.38×10^{-23} J/K), T is the temperature in Kelvin (300 K in this case), R is the resistance in ohms, and Δf is the bandwidth in Hz, set in this experiment to 1 Hz (typically the FFT line width).

Figure 5 shows the *PSD* distribution and noise contributions in the 3 microbolometers with different a-SiGe:H,N thermo-sensing films, where it is observed that the microbolometer with film #1 has the lowest V_n value. The results also show that $1/f$ noise predominates at low frequencies, while thermal noise becomes more significant at higher frequencies. Table 2 presents the noise voltage (Δv) in terms of the square root of frequency, which provides information on the sensor thermal noise level. A lower Δv indicates less noise in the sensor, which is generally desirable as it implies a cleaner signal [22]. Also, Table 2 shows the room temperature resistance R (Ω) which was experimentally measured.

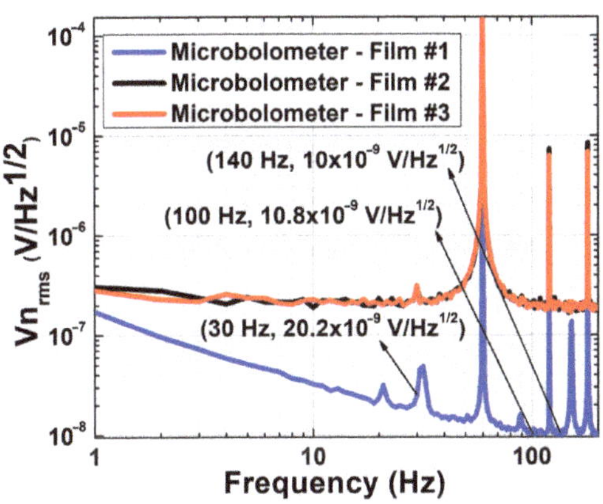

Figure 5. Power Spectral Density (*PSD*) distribution and noise contributions in 3 microbolometers with different a-SiGe:H,N thermo-sensing films.

Table 2. Noise, room temperature resistance, *NEP*, detectivity and *NETD* values of microbolometers with three different a-SiGe:H,N thermo-sensing films.

Microbolometer Film	Frequency (Hz)	v_n $\left({}^{\text{Vrms}}/{}_{\sqrt{\text{Hz}}}\right)$	Room Temperature Resistance (Ω)	*NEP* $\left({}^{\text{W}}/{}_{\sqrt{\text{Hz}}}\right)$	Detectivity $D^*\left({}^{\text{cm}\sqrt{\text{Hz}}}/{}_{\text{W}}\right)$	*NETD* (mK)
Film #1	30	20.2×10^{-9}	2.76×10^3	2.78×10^{-14}	1.80×10^9	45.15
	100	10.8×10^{-9}		1.49×10^{-14}	3.36×10^9	24.14
	140	10.0×10^{-9}		1.38×10^{-14}	3.63×10^9	22.35
Film #2	30	313×10^{-9}	15.7×10^9	6.28×10^{-13}	7.96×10^7	1018
	100	196×10^{-9}		3.92×10^{-13}	1.28×10^8	637.38
	140	198×10^{-9}		3.96×10^{-13}	1.26×10^8	643.88
Film #3	30	314×10^{-9}	5.72×10^6	3.84×10^{-13}	1.30×10^8	624.43
	100	194×10^{-9}		2.38×10^{-13}	2.10×10^8	186.93
	140	180×10^{-9}		2.20×10^{-13}	2.27×10^8	357.95

2.5. Noise Equivalent Power (NEP)

The *NEP* represents the minimum detectable power of the microbolometer, where the signal is equal to the noise level, the equation used is

$$NEP = \frac{v_n}{R_v}\left({}^{\mathrm{W}}/_{\sqrt{\mathrm{Hz}}}\right) \tag{4}$$

where v_n is the measured noise voltage and R_v is the voltage responsivity.

Table 2 shows the *NEP* values of the three different microbolometers with different a-SiGe:H,N thermos-sensing films for three different values of frequency. The tests yielded *NEP* values in the range of 1×10^{-14} $(\mathrm{W}/\sqrt{\mathrm{Hz}})$, demonstrating the high sensitivity of the device [23].

In the evaluation of the microbolometers with the three different thermos-sensing films, film #1 has proven to be the most efficient option in terms of sensitivity and stability. This configuration presents an optimal balance between low noise levels and high responsivity, resulting in an improved *NEP*. Additionally, the resistance values in this film are moderate, ensuring effective thermal response without compromising operational stability. These factors make this combination ideal for high-precision infrared detection applications, maximizing the signal-to-noise ratio and optimizing the overall performance of the device [24].

2.6. Detectivity (D)*

The normalized detectivity (D^*) indicates the ability of the microbolometer to detect weak signals in the presence of noise. The equation used is

$$D^* = \frac{\Re_V\sqrt{A\Delta f}}{V_n}\left[\frac{\mathrm{cm}\sqrt{\mathrm{Hz}}}{\mathrm{W}}\right] \tag{5}$$

where A is the detector area in cm^2, Δf is the bandwidth, $\Re_V$ is the voltage responsivity, and V_n is the noise voltage, previously calculated [25].

For the detectivity calculation, tests were performed under controlled temperature and frequency conditions. Equation (5) was used, considering the microbolometer area of 50 μm × 50 μm. The results are shown in Table 2, indicating that the highest detectivity corresponds to the microbolometer with film #1. As the frequency increased to 30 Hz, 100 Hz, and 140 Hz, the detectivity values were still high, suggesting that this microbolometer maintains high sensitivity across the frequency spectrum. This highlights the microbolometers ability to differentiate between incident radiation signals and background noise. The higher the detectivity, the greater the detector sensitivity, enabling it to detect weaker signals above the noise level [26]. A detectivity of the order of 10^9 $\mathrm{cm}\sqrt{\mathrm{Hz}}/\mathrm{W}$ was obtained, which is suitable for high-precision thermal imaging applications.

2.7. Noise Equivalent Temperature Difference (NETD)

The *NETD* represents the minimum temperature difference that the microbolometer can detect, indicating its thermal sensitivity. The equation used to calculate the *NETD* is

$$NETD = \frac{4F^2V_n}{\tau_0 A\Re_V\left(\frac{\Delta P}{\Delta T}\right)_{\lambda_1-\lambda_2}} \tag{6}$$

where $\left(\frac{\Delta P}{\Delta T}\right)_{\lambda_1-\lambda_2}$ is the power emitted by a blackbody per unit of temperature within the spectral range, A is the area of the microbolometer, τ_0 is the transmittance of the optical system, F is the f-number of the lens, V_n is the measured noise voltage, and $\Re_V$ is the responsivity of the device [27].

The *NETD* tests were conducted using an F/1 optical system with a transmittance $\tau_0 = 0.5$, along with a temperature contrast in the 8–12 μm spectral band at 300 K of $\left(\frac{\Delta P}{\Delta T}\right)_{8-12} = 1.972\ \mathrm{W/m^2K}$. Table 2, shows the *NETD* values calculated at three different frequencies.

The microbolometer demonstrated a *NETD* below 50 mK, ensuring its ability to detect very subtle temperature changes in the range of 8 to 12 μm of the infrared spectrum. The obtained *NETD* values confirm the high thermal sensitivity of the microbolometer, with film #1 being the most suitable for high-precision thermal imaging applications.

2.8. Resistance Versus Temperature (R(T))

Finally, we performed measurement of the microbolometer resistance as a function of temperature *R(T)* to demonstrate its operational range. For this characterization, the microbolometer was placed in a vacuum cryostat (MMR Technologies, San Jose, CA, USA) at 40 mTorr and current-voltage I(V) characteristics were obtained at different temperatures, in a range of 300 K to 430 K, with steps of 10 K.

Figure 6a shows the voltage bias plotted as a function of current from the microbolometer with the a-SiGe:H,N film #1. For clarity, the selected curves are shown at temperatures of 300 K, 350 K, 370 K, 410 K, and 430 K. In each curve, in the linear region, the slope was extracted, corresponding to the electrical resistance.

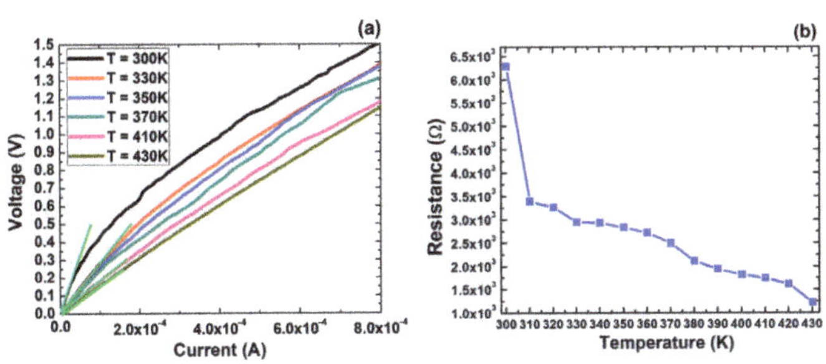

Figure 6. (**a**) Current-voltage I(V) characteristics of the microbolometer with a-SiGe:H,N film #1 at different temperatures ranging from 300 K to 430 K. (**b**) Temperature dependence of the electrical resistance R(T) of the microbolometer with a-SiGe:H,N film #1.

From these measurements, the temperature dependence of the electrical resistance *R(T)* was obtained and is presented in Figure 6b. The resulting curve reveals a clear negative temperature coefficient of resistance (*TCR*), characteristic of amorphous semiconductors. This behavior confirms that the film remains sensitive throughout the measured range, making it suitable for uncooled infrared detection applications.

3. Materials and Methods

3.1. Thermo-Sensing Film Deposition Conditions

Three different a-SiGe:H,N films were used as thermos-sensing films in microbolometers, based on our previous work, where these kinds of films were studied [28]. The films were deposited from an SiH_4, GeH_4, N_2, and H_2 gas mixture in a low frequency (110 KHz) capacitively coupled PECVD reactor (Applied Materials, Inc., Santa Clara, CA, USA) at 200 °C, with an RF power of 300 W (corresponding to a power density of 87 mW/cm^2). Film #1 was deposited with flow rates of SiH_4 = 40 sccm/GeH_4 = 10 sccm/N_2 = 100 sccm/

H_2 = 1000 sccm. This film exhibits an activation energy (E_a) of 0.36 eV, a *TCR* of 4.65%/K, and conductivity of 7.56 × 10^{-6} (Ω·cm)$^{-1}$. Film #2 was deposited with flow rates of SiH_4 = 30 sccm/GeH_4 = 20 sccm/N_2 = 100 sccm/H_2 = 1000 sccm, with an E_a of 0.24 eV, a *TCR* of 3.15%/K, and conductivity of 1.47 × 10^{-3} (Ω·cm)$^{-1}$. Finally, film #3 was deposited with flow rates of SiH_4 = 10 sccm/GeH_4 = 40 sccm/N_2 = 100 sccm/H_2 = 1000 sccm. This film has an E_a of 0.17 eV and a low *TCR* of 2.23%/K but a high conductivity (3.24 × 10^{-2} (Ω·cm)$^{-1}$), which facilitates its integration with the read-out integrated circuit (ROIC) based on the silicon CMOS technology [29].

3.2. Microbolometers Fabrication

The fabrication of microbolometers was carried out using silicon technology on a two-inch silicon wafer. The microbolometer fabrication process began with the initial silicon wafer surface cleaning using the standard RCA process: RCA I composed of deionized water/ammonium hydroxide/hydrogen peroxide and RCA II composed of deionized water/hydrogen chloride/hydrogen peroxide (J.T. Baker, Easton, PA, USA). The above cleaning process is used to remove organic and metallic contaminants. The native oxide of the wafer was removed using a solution of deionized (DI) water and hydrofluoric acid (J.T. Baker, Easton, PA, USA), preparing the wafer for subsequent stages (Figure 7a). A 1 µm thick layer of silicon dioxide (SiO_2) was then grown across the wafer, providing thermal and electrical insulation between the microbolometer and the substrate (Figure 7b) [30]. A 0.6 µm titanium layer was deposited using electron beam evaporation (e-beam). The lift-off technique was used to define the metallic contacts and reflective mirrors [31]. These mirrors redirect unabsorbed infrared radiation back toward the sensing film, maximizing detection efficiency (Figure 7c).

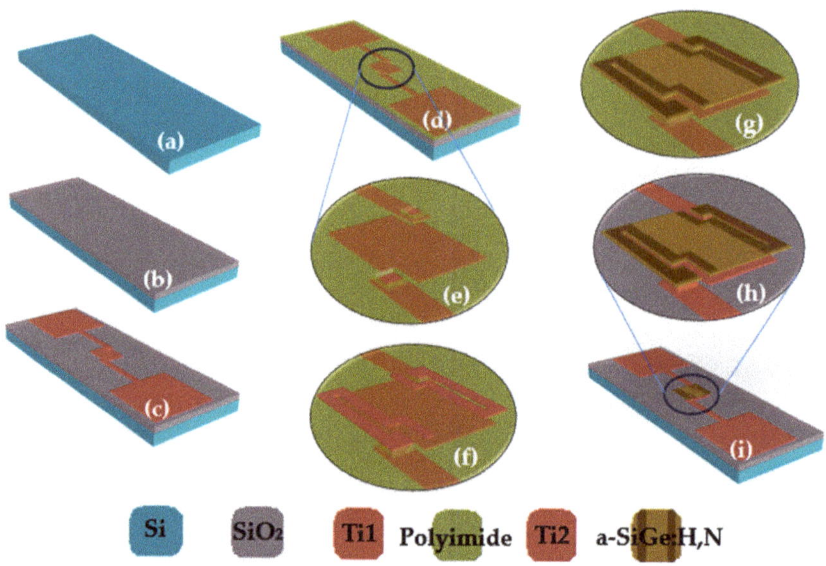

Figure 7. Fabrication process scheme of the microbolometer: (**a**) Cleaning of the silicon wafer, (**b**) SiO_2 deposition for thermal insulation, (**c**) titanium deposition and patterning for contacts and reflective mirrors, (**d**) Application of the sacrificial polyimide layer, (**e**) Etching to expose the contacts, (**f**) Titanium deposition and patterning for sensor film contact electrodes, (**g**) Deposition and definition of the a-SiGe:H,N film, (**h**) Complete removal of the polyimide sacrificial film, (**i**) final fabricated microbolometer.

To form the microbolometer, the goal is to thermally insulate a sensor film by suspending it from the surface. A sacrificial polyimide 2610 Kapton film (DuPont corporation, Wilmington, DE, USA), with a thickness of 2.5 μm, was applied across the surface using the spin coating technique at 2500 rpm. This creates a Fabry-Perot resonant cavity at a quarter wavelength of 10 μm, corresponding to 300 K, enhancing absorption and sensitivity (Figure 7d) [32]. Windows were etched in the polyimide layer to expose the underlying metal contacts. For this step, a 120 nm-thick aluminum film was used as a hard mask, patterned by the lift-off process. A photomask with a dark-field layout and positive photoresist AZ1505 (MicroChemicals GmbH, Ulm, Germany) was employed. The pattern was transferred to the polyimide using anisotropic etching via a Reactive Ion etching, RIE AME-8110 tool (Applied Materials, Inc., Santa Clara, CA, USA). The etching process was carried out using an oxygen plasma at a pressure of 100 mTorr, RF power of 250 W, and gas flow rate of 80 sccm, with a total etching time of 20 min. These parameters were optimized to ensure clean and accurate pattern transfer, enabling proper electrical continuity through the exposed contacts. (Figure 7e).

Subsequently, an additional 0.4 μm titanium layer was deposited to form the connection electrodes between the metal pads and the sensor film. These electrode patterns were defined using lithography and a lift-off process (Figure 7f). Over the electrodes, a 0.5 μm thick a-SiGe:H,N was deposited, acting as the infrared-absorbing film (Figure 7g). The addition of N_2 to the film enhances the infrared absorption, as was reported in our previous work [28]. The films deposition was performed via low frequency PECVD (Applied Materials, Inc., Santa Clara, CA, USA) working at 110 KHz, with an RF power of 300 W (corresponding to a power density of 87 mW/cm^2) and a substrate temperature of 200 °C. After deposition, thermal treatment was conducted in the chamber without breaking the vacuum, allowing a nitrogen flow of 20 sccm, a pressure of 1200 mTorr, and a temperature of 200 °C. This process improves the films residual stress and optimizes its conductivity [33]. The treatment lasted for 4 continuous hours.

The process concluded with the removal of the sacrificial polyimide film using oxygen plasma with a barrel asher L2101 (Branson/IPC, Hayward, CA, USA) for isotropic etching. The above allows the release of the suspended membrane, providing thermal insulation to the sensing film and improving the response to rapid temperature variations induced by infrared (IR) radiation (Figure 7h). The device, with dimensions of 50 × 50 μm^2, has low thermal capacity, ensuring superior performance in infrared detection applications by minimizing heat loss and enhancing sensitivity (Figure 7i). Figure 8 shows images of the fabricated microbolometer obtained with a scanning electron microscope (SEM) SU3500 (Hitachi, Tokyo, Japan).

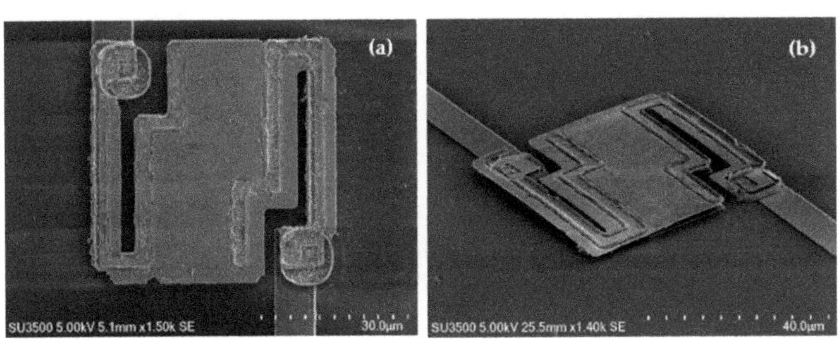

Figure 8. SEM images of the fabricated microbolometer of an area of 50 μm × 50 μm: (**a**) top view and (**b**) view at a 60 degree inclination.

4. Conclusions

The research carried out in this work demonstrated that microbolometers fabricated with nitrogen-doped hydrogenated silicon-germanium films (a-SiGe:H,N), using LF-PECVD, achieved excellent thermal sensitivity and operational stability. Particularly the device configuration with the a-SiGe:H,N film deposited with a ratio of SiH_4/GeH_4 = 40/10 (highly diluted in H_2 and N_2), reached a voltage responsivity of 7.26×10^5 V/W and a *NETD* of 22.35 mK at 140 Hz. These characteristics and a significant reduction in 1/f noise above 100 Hz highlight the devices potential for high-precision uncooled thermal imaging applications, making it compatible with silicon CMOS technology and suitable for integration into portable and low-power systems.

Author Contributions: Conceptualization, O.V., A.T., R.J. and M.M.; methodology, O.V., A.T., L.H. and M.M.; validation, O.V., R.J., A.M., L.H., A.L. and K.M.; resources, A.L., J.F. and G.M.M.; writing—original draft preparation, O.V., R.J., A.T., L.H., M.M. and A.M.; writing—review and editing, O.V., A.L., K.M., M.M., J.F. and G.M.M.; visualization, O.V., A.M. and M.M.; supervision, R.J., A.T., A.M., A.L. and M.M.; project administration, O.V., L.H., K.M., J.F. and G.M.M.; funding acquisition, M.M., A.L., J.F. and G.M.M. All authors have read and agreed to the published version of the manuscript.

Funding: This research received no external funding.

Data Availability Statement: The original contributions presented in this study are included in the article. Further inquiries can be directed to the corresponding author (Any inquiry can be directed to mmoreno@inaoep.mx).

Acknowledgments: O. Velandia acknowledges SECIHTI for the scholarship 2020-000026-02NACF-07348 granted during the Ph.D. program at INAOE. M. Moreno acknowledges SECIHTI for the sabbatical stay support.

Conflicts of Interest: The authors declare no conflicts of interest.

References

1. Hornberger, C. Application of Military Uncooled Infrared Sensors to Homeland Defense. *Proc. SPIE* **2002**, *4708*, 201–211. [CrossRef]
2. Lahiri, B.B.; Bagavathiappan, S.; Jayakumar, T.; Philip, J. Medical Applications of Infrared Thermography: A Review. *Infrared Phys. Technol.* **2012**, *55*, 221–235. [CrossRef]
3. Tissot, J.-L.; Trouilleau, C.; Fieque, B.; Crastes, A.; Legras, O. Uncooled Microbolometer Detector: Recent Developments at ULIS. *Opto-Electron. Rev.* **2006**, *14*, 25–32. [CrossRef]
4. Yon, J.J.; Mottin, E.; Biancardini, L.; Letellier, L.; Tissot, J.L. Infrared Microbolometer Sensors and Their Application in Automotive Safety. *Adv. Microsyst. Automot. Appl.* **2003**, *13*, 137–157. [CrossRef]
5. Dufour, D.; Le Noc, L.; Tremblay, B.; Tremblay, M.N.; Généreux, F.; Terroux, M.; Vachon, C.; Wheatley, M.J.; Johnston, J.M.; Wotton, M.; et al. A Bi-Spectral Microbolometer Sensor for Wildfire Measurement. *Sensors* **2021**, *21*, 3690. [CrossRef] [PubMed]
6. Kusnierek, K.; Korsaeth, A. Challenges in Using an Analog Uncooled Microbolometer Thermal Camera to Measure Crop Temperature. *Int. J. Agric. Biol. Eng.* **2014**, *7*, 60–74. [CrossRef]
7. Yu, L.; Guo, Y.; Zhu, H.; Luo, M.; Han, P.; Ji, X. Low-cost microbolometer type infrared detectors. *Micromachines* **2020**, *11*, 800. [CrossRef]
8. Shin, C.; Kim, J.-K.; Oh, J.; Park, J.; Shin, H.-S. Noise Improvement of a-Si Microbolometers by the Post-Metal Annealing Process. *Sensors* **2021**, *21*, 7103. [CrossRef]
9. Almasri, M.; Xu, B.; Castracane, J. Amorphous Silicon Two-Color Microbolometer for Uncooled IR Detection. *IEEE Sens. J.* **2006**, *6*, 293–300. [CrossRef]
10. Piller, M.; Hiesberger, J.; Wistrela, E.; Martini, P.; Luhmann, N.; Schmid, S. Thermal IR Detection With Nanoelectromechanical Silicon Nitride Trampoline Resonators. *IEEE Sens. J.* **2023**, *23*, 1066–1071. [CrossRef]
11. Hu, Y.; Lin, C.H.; Min, S.; Smith, R.L.; Roberts, S. Bolometric Arrays and Infrared Sensitivity of VO_2 Films with Varying Stoichiometry. *J. Optoelectron. Mater. Eng.* **2017**, *8*, 345–353.

12. Shin, C.; Pham, D.P.; Park, J.; Kim, S.; Lee, Y.-J.; Yi, J. Structure and Electrical Properties of Boron Doped Hydrogenated Mixed-Phase Silicon Films for Uncooled Microbolometer. *Infrared Phys. Technol.* **2018**, *94*, 103–111. [CrossRef]
13. Kolahdouz, M.; Salemi, A.; Moeen, M.; Östling, M.; Radamson, H.H. Kinetic Modeling of Low Temperature Epitaxy Growth of SiGe Using Disilane and Digermane. *J. Electrochem. Soc.* **2012**, *159*, H478–H481. [CrossRef]
14. Jung, J.; Kim, M.; Kim, C.-H.; Kim, T.H.; Park, S.H.; Kim, K.; Cho, H.J.; Kim, Y.; Kim, H.Y.; Oh, J.S. Improved responsivity of an a-Si-based micro-bolometer focal plane array with a SiNx membrane layer. *J. Sens. Sci. Technol.* **2022**, *31*, 366–370. [CrossRef]
15. Forsberg, F.; Lapadatu, A.; Kittilsland, G.; Martinsen, S.; Roxhed, N.; Fischer, A.C.; Stemme, G.; Samel, B.; Ericsson, P.; Høivik, N.; et al. CMOS-integrated Si/SiGe quantum-well infrared microbolometer focal plane arrays manufactured with very large-scale heterogeneous 3-D integration. *IEEE J. Sel. Top. Quantum Electron.* **2015**, *21*, 2700111. [CrossRef]
16. Moreno, M.; Jiménez, R.; Torres, A.; Ambrosio, R. Microbolometers based on amorphous silicon–germanium films with embedded nanocrystals. *IEEE Trans. Electron Devices* **2015**, *62*, 2120–2127. [CrossRef]
17. Rogalski, A. *Infrared Detectors*; Gordon and Breach Science Publishers: Amsterdam, The Netherlands, 2000.
18. Budzier, H.; Gerlach, G. *Thermal Infrared Sensors: Theory, Optimisation and Practice*; Wiley-VCH: Weinheim, Germany, 2011.
19. Jimenez, R.; Moreno, M.; Torres, A.; Morales, A.; Ponce, A.; Ferrusca, D.; Rangel-Magdaleno, J.; Castro-Ramos, J.; Hernandez-Perez, J.; Cano, E. Fabrication of Microbolometer Arrays Based on Polymorphous Silicon–Germanium. *Sensors* **2020**, *20*, 2716. [CrossRef]
20. Xu, Q.; Zhou, Z.; Tan, C.; Pan, X.; Wen, Z.; Zhang, J.; Zhou, D.; Sun, Y.; Chen, X.; Zhou, L.; et al. A Spectrally Selective Visible Microbolometer Based on Planar Subwavelength Thin Films. *Nanoscale Adv.* **2023**, *5*, 2054–2060. [CrossRef]
21. Hai, M.L.; Cheng, Q.; Hesan, M.; Qu, C.; Kinzel, E.C.; Almasri, M. Amorphous $Si_xGe_yO_{1-x-y}$ Thin Films for Uncooled Infrared Microbolometers. *Infrared Phys. Technol.* **2018**, *95*, 227–235. [CrossRef]
22. Gao, Y.; Chen, H.; Xu, Y.; Sun, X.; Chang, B. Noise Research of Micro-Bolometer Array under Temperature Environment. *Proc. SPIE Int. Symp. Photoelectron. Detect. Imaging* **2011**, *8193*, 81930S. [CrossRef]
23. Bartmann, M.G.; Sistani, M.; Luhmann, N.; Schmid, S.; Bertagnolli, E.; Lugstein, A.; Smoliner, J. Germanium Nanowire Microbolometer. *Nanotechnology* **2022**, *33*, 245201. [CrossRef] [PubMed]
24. Cosme Bolaños, I. Experimental Study of Noise in Thermo-Sensing Films Based on Silicon-Germanium Deposited by Plasma for Application in Microbolometers. Master's Thesis, National Institute of Astrophysics, Optics and Electronics (INAOE), Tonantzintla, Puebla, Mexico, 2008.
25. Abdullah, A.; Koppula, A.; Alkorjia, O.; Almasri, M. Uncooled Two-Microbolometer Stack for Long Wavelength Infrared Detection. *Sci. Rep.* **2023**, *13*, 3470. [CrossRef] [PubMed]
26. Adiyan, U.; Larsen, T.; Zárate, J.J.; Villanueva, L.G.; Shea, H. Shape Memory Polymer Resonators as Highly Sensitive Uncooled Infrared Detectors. *Nat. Commun.* **2019**, *10*, 4518. [CrossRef] [PubMed]
27. Perić, D.; Livada, B.; Perić, M.; Vujić, S. Thermal Imager Range: Predictions, Expectations, and Reality. *Sensors* **2019**, *19*, 3313. [CrossRef]
28. Velandia, O.; Moreno, M.; Zavala, R. Optimization of the Electrical Conductivity and Thermal Coefficient of Resistance (TCR) on Hydrogenated Amorphous Silicon–Germanium Films Doped with Nitrogen (a-SiGe:H,N) for Applications on High-Performance Infrared Detectors. In Proceedings of the 2023 IEEE Latin American Electron Devices Conference (LAEDC), Puebla, Mexico, 3–5 July 2023. [CrossRef]
29. Chu, P.S.; Chen, H.L.; Ding, R.J. A Novel Linear-Logarithmic Readout Integrated Circuit with High Dynamic Range. *Infrared Phys. Technol.* **2019**, *102*, 103158. [CrossRef]
30. Calleja Gómez, C.M. Fabrication and Characterization of UV Radiation Detectors Based on Silicon. Master's Thesis, National Institute of Astrophysics, Optics and Electronics (INAOE), Tonantzintla, Puebla, Mexico, 2011.
31. Alanís, A.; Díaz, D.; Reyes, C.; Zúñiga, C.; Torres, A.; Rosales, P.; Molina, J.; Hidalga, J.; Linares, M.; Aceves, M.; et al. PolyMEMS INAOE, a Surface Micromachining Fabrication Module and the Development of Microstructures for Residual Stress Analysis. In Proceedings of the 6th Ibero-American Congress on Sensor, IBERSENSORS, Sao Paulo, Brazil, 24–28 November 2008. Available online: https://www.researchgate.net/publication/312498987_PolyMEMS_INAOE_a_Surface_Micromachining_Fabrication_Module_and_the_Development_of_Microstructures_for_Residual_Stress_Analysis (accessed on 1 September 2024).
32. Yan, W.; Shrestha, V.R.; Jeangros, Q.; Sefidmooye Azar, N.; Balendhran, S.; Ballif, C.; Crozier, K.B.; Bullock, J. Spectrally Selective Mid-Wave Infrared Detection Using Fabry–Pérot Cavity Enhanced Black Phosphorus 2D Photodiodes. *ACS Nano* **2020**, *14*, 13547–13556. [CrossRef]
33. Abdel-Rahman, M.; Zia, M.; Alduraibi, M. Temperature-Dependent Resistive Properties of Vanadium Pentoxide/Vanadium Multi-Layer Thin Films for Microbolometer & Antenna-Coupled Microbolometer Applications. *Sensors* **2019**, *19*, 1320. [CrossRef]

inorganics

MDPI

Article

Enhanced Antimicrobial and Biomedical Properties of Fe-Based Bulk Metallic Glasses Through Ag Addition

Long Jiang [1,2], Xueru Fan [1,2], Qiang Li [1,2,*], Xin Li [1,2], Tao Jiang [3,4,*] and Qin Wei [5]

1 School of Materials Science and Engineering, Xinjiang University, Urumqi 830046, China; tsyjiangl0924@163.com (L.J.); 18835561709@163.com (X.F.); lx1272636024@163.com (X.L.)
2 Xinjiang Environmental Functional Materials Engineering Technology Research Center, Xinjiang University, Urumqi 830017, China
3 Animal Laboratory Center, Xinjiang Medical University, Urumqi 830054, China
4 Xinjiang Key Laboratory of Medical Animal Model Research, Urumqi 830054, China
5 Central Laboratory, Xinjiang Medical University, Urumqi 830054, China; weiqin1027@126.com
* Correspondence: qli@xju.edu.cn (Q.L.); jtao0215@126.com (T.J.)

Abstract: This study explores the enhancement of antimicrobial and biomedical properties in Fe-based bulk metallic glasses (BMGs) through the addition of Ag. $Fe_{55-x}Cr_{20}Mo_5P_{13}C_7Ag_x$ (x = 0, 1, 2, 3 at.%) master alloy ingots were synthesized by the induction melting technique and industrial-grade raw materials, the master alloy ingots were prepared as bulk metallic glasses (referred to as Ag0, Ag1, Ag2, and Ag3) by the water-cooled copper-mold suction casting technique, and their glass-forming ability, corrosion resistance, biocompatibility, and antimicrobial properties were systematically investigated. The results indicate that the glass forming ability (GFA) decreased with increasing Ag content, reducing the critical diameter for fully amorphous formation from 2.0 mm for Ag0 to 1.0 mm for Ag3. Electrochemical tests in Hank's solution revealed the superior corrosion resistance of the Fe-based BMGs as compared with conventional 316 L stainless steel (316L SS) and Ti6Al4V alloy (TC4), with Ag3 demonstrating the lowest corrosion current density and the most stable passivation. Biocompatibility assessments, including fibroblast cell viability and adhesion tests, showed enhanced cellular activity and morphology on Fe-based BMG surfaces as compared with 316L SS and TC4, with minimal harmful ion release. Antimicrobial tests against *E. coli* and *S. aureus* revealed significantly improved performance with the Ag addition, achieving bacterial inhibition rates of up to 87.5% and 86.7%, respectively, attributed to Ag^+-induced reactive oxygen species (ROS) production. With their excellent corrosion resistance, biocompatibility, and antimicrobial activity, the present Ag-containing Fe-based BMGs, particularly Ag3, are promising candidates for next-generation biomedical implants.

Keywords: Fe-based bulk metallic glasses; Ag addition; antimicrobial properties; biocompatibility

1. Introduction

The continuous development of the medical field has driven the demand for enhanced biomedical materials with superior functionality, safety, and patient comfort. Currently, ceramics, medical stainless steel (316L SS), and titanium alloys (e.g., Ti6Al4V, TC4) dominate the medical implant market due to their mechanical strength, corrosion resistance, and biocompatibility. However, these materials have notable limitations that restrict their application in modern medical practices. For instance, 316L SS is prone to corrosion in physiological environments, leading to the release of Ni ions, which can cause inflammatory and allergic reactions such as nickel allergy [1]. Additionally, its wear resistance and antimicrobial properties are insufficient for advanced medical applications. Although TC4 titanium alloy offers better corrosion resistance and biocompatibility, its inert surface often necessitates additional activation treatments to improve cell adhesion and reduce bacterial colonization [2,3].

One of the most critical challenges of using traditional materials is the high risk of device-associated infections primarily caused by bacterial adhesion and biofilm formation on implant surfaces [4]. Such infections contribute to prolonged patient recovery, high morbidity, and increased healthcare costs. The global rise of drug-resistant and multidrug-resistant bacteria has further amplified this issue, necessitating the development of novel antimicrobial biomaterials [5–7]. Conventional disinfection and sterilization techniques often fail to eradicate biofilms, prompting the exploration of advanced materials capable of intrinsic antimicrobial activity.

Recent advancements in materials science have highlighted the potential of bulk metallic glasses (BMGs) as next-generation biomaterials. BMGs are unique metallic alloys with amorphous structures characterized by the absence of grains, grain boundaries, and dislocations. This structure endows BMGs with exceptional mechanical properties, superior corrosion resistance, and improved biocompatibility as compared with traditional crystalline metals [8,9]. Thus, BMGs have a good prospect for biomedical use and potentially as biomedical materials, especially as surgical instruments such as gastrointestinal endoscopes, razor blades, and other medical devices such as artificial implants [8,10]. In particular, Fe-based BMGs have emerged as promising candidates for biomedical applications due to their cost-effectiveness and compatibility with magnetic resonance imaging (MRI), outperforming alternatives like 316L SS [11,12]. However, despite their outstanding corrosion resistance and mechanical strength, the antimicrobial properties of Fe-based BMGs remain underexplored.

Among the various strategies for imparting antimicrobial properties, the incorporation of Ag into biomaterials has gained significant attention. Ag is widely recognized as a potent antimicrobial agent due to its ability to disrupt bacterial membranes and metabolic pathways through the generation of reactive oxygen species (ROS) [13]. Unlike antibiotics, Ag remains effective against multidrug-resistant bacteria, making it an ideal additive for biomedical applications [14–19]. Incorporating Ag into Fe-based BMGs could potentially enhance their antimicrobial performance while retaining the intrinsic advantages of BMGs, such as high strength and biocompatibility.

This study seeks to address the gap in the development of antimicrobial Fe-based BMGs by examining the effects of Ag addition on their structural, thermal, mechanical, and electrochemical properties, as well as their biocompatibility and antimicrobial efficacy. Previous work by Li et al. demonstrated the development of $Fe_{55}Cr_{20}Mo_5P_{13}C_7$ BMG as a novel biomedical implant material, showcasing excellent corrosion resistance and biocompatibility [3]. In this work, Ag is incorporated into this Fe-based BMG to further enhance its antibacterial properties. The antimicrobial effectiveness of these Ag-containing Fe-based BMGs is evaluated using both Gram-negative bacteria (*E. coli*) and Gram-positive bacteria (*S. aureus*), pathogens commonly implicated in healthcare-associated infections. Comparative studies with 316L SS and TC4 alloys are also conducted to benchmark the performance of the developed materials.

This work not only demonstrates the enhanced antimicrobial properties of Fe-based BMGs through Ag addition but also highlights their potential as cost-effective, biocompatible materials for biomedical implants. By addressing the dual challenges of bacterial infection and material biocompatibility, this study contributes to advancing the field of medical materials and presents Fe-based BMGs as promising candidates for next-generation implantable devices.

2. Results and Discussion

2.1. Structural, Thermal, and Mechanical Analyses

The X-Ray diffraction (XRD) patterns of the as-cast $Fe_{55-x}Cr_{20}Mo_5P_{13}C_7Ag_x$ (x = 0, 1, 2, 3 at.%) samples, shown in Figure 1, reveal the critical diameter for fully amorphous formation (D_{max}). All samples exhibit a broad diffuse peak near the diffraction angle 2θ = 45°, with no discernible sharp-crystallization peaks. This confirms that the Fe-based bulk metallic glasses (BMGs) prepared in this study possess a fully amorphous structure, free from any crystalline phases. Notably, the absence of toxic or rare earth elements further underscores the suitability of these alloys for biomedical applications. For simplicity in the subsequent discussion, the $Fe_{55}Cr_{20}Mo_5P_{13}C_7Ag_0$, $Fe_{54}Cr_{20}Mo_5P_{13}C_7Ag_1$, $Fe_{53}Cr_{20}Mo_5P_{13}C_7Ag_2$, and $Fe_{52}Cr_{20}Mo_5P_{13}C_7Ag_3$ BMGs will be referred to Ag0, Ag1, Ag2, and Ag3, respectively.

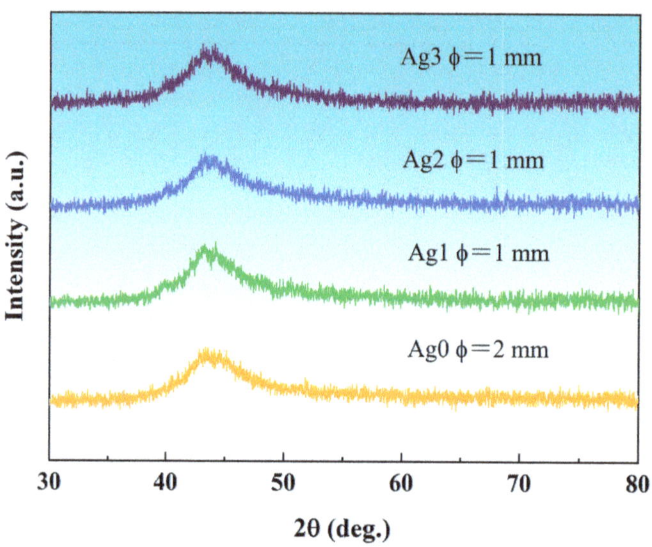

Figure 1. XRD patterns of the as-cast $Fe_{55-x}Cr_{20}Mo_5P_{13}C_7Ag_x$ (x = 0, 1, 2, 3 at.%) glassy alloy rod samples with critical diameter for fully amorphous formation.

Figure 2 presents the DSC curves of the as-cast Fe-based BMG samples at a heating rate of 0.33 K/s. All samples exhibit a pronounced glass transition, followed by a supercooled liquid-phase region and a multi-step crystallization process [3]. These thermal features are indicative of their amorphous natures and provide a basis for determining essential thermodynamic parameters, such as glass transition temperature (T_g), onset crystallization temperature (T_x), melting temperature (T_m), and liquidus temperature (T_l), which are summarized in Table 1. The analysis reveals that both T_g and T_x increase with the addition of Ag content in the Fe-based BMGs.

To investigate the effect of Ag content on the glass-forming ability (GFA) of the Fe-based alloys, the three most commonly used GFA criteria, i.e., the reduced glass transition temperature T_{rg} (=T_g/T_l), the γ parameter (=$T_x/(T_g + T_l)$) and the supercooled liquid-phase region ΔT_x (=$T_x - T_g$), were calculated and are listed in Table 1. These results show that the T_{rg} and γ values remain relatively consistent at approximately 0.58 and 0.38, respectively, across all samples. However, ΔT_x decreases as Ag content increases, indicating a reduction in GFA with higher Ag content. This trend aligns with the experimentally determined results of D_{max}. The observed reduction in GFA can be attributed to the enthalpy of mixing between the Ag and Fe atoms, which is calculated as +28 kJ/mol [20]. This positive enthalpy value indicates a repulsive interaction between Ag and Fe atoms. Such interactions promote the precipitation of the α-Fe phase, thereby disrupting the amorphous structure and significantly reducing the GFA of Fe-based BMGs. This behavior is analogous to the effect observed with the addition of Cu in Fe-based amorphous alloys [21], where similarly repulsive atomic interactions favor the nanocrystallization of α-Fe and hinder glass formation.

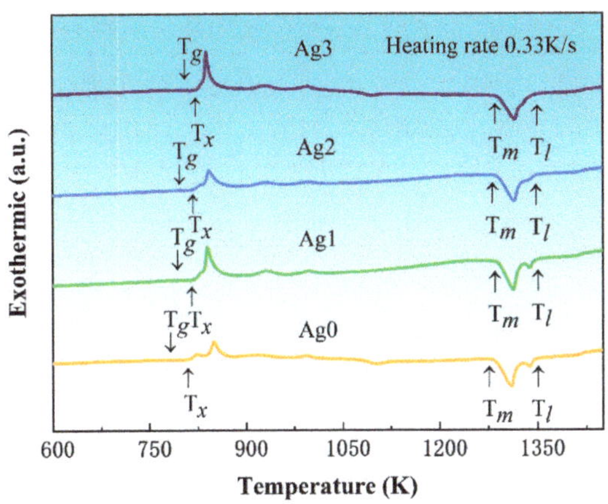

Figure 2. DSC thermal scans of the Fe-based BMGs at a heating rate of 0.33 K/s.

Table 1. Thermodynamic parameters and GFA criterion values for the Fe-based BMGs samples (T_m: melting temperature; T_l: liquidus temperature; T_g: glass transition temperature; reduced glass transition temperature: $T_{rg} = T_g/T_l$; width of supercooled liquidus zone: $\Delta T = T_x - T_g$; γ criterion: $\gamma = T_x/T_g + T_l$).

Samples	D_{max} (mm)	T_g (K)	T_x (K)	T_m (K)	T_l (K)	T_{rg}	ΔT_x (K)	γ
Ag0	2	784	809	1280	1349	0.58	25	0.380
Ag1	1	794	813	1284	1348	0.58	19	0.379
Ag2	1	797	815	1282	1349	0.59	18	0.379
Ag3	1	805	820	1286	1396	0.57	15	0.372

Figure 3 shows the compression curves of Fe-based BMGs at room temperature, and the test samples are all 1 mm in diameter and 2 mm in length. All the samples exhibit brittle fracture after linear elastic deformation. The Fe-based BMGs exhibit ultrahigh compression strength (σ_f) exceeding 2 GPa, which gradually increases with the increase in Ag content and reaches the largest value of 2.34 GPa for the Ag3. It is well known that the strength of amorphous alloys is proportional to their T_g [22]. It can be seen from Table 1 that the variation in σ_f of the Fe-based BMGs is consistent with their T_g values, showing a positive correlation.

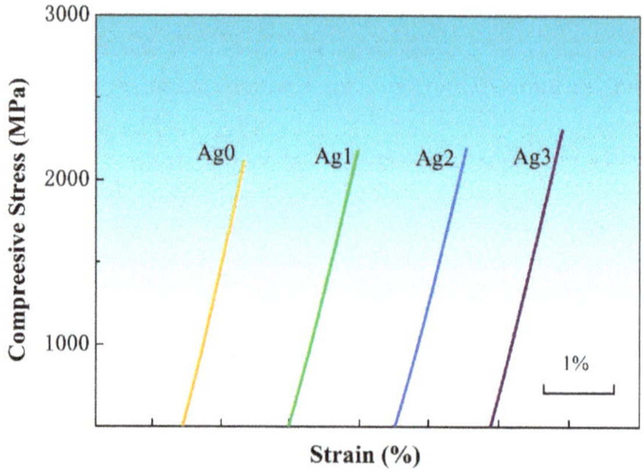

Figure 3. Engineering stress–strain curves of the as-cast Fe-based amorphous alloy rod samples with a diameter of 1 mm and a length of 2 mm at room temperature.

2.2. Electrochemical Measurement

Figure 4 presents the potentiodynamic polarization curves of Fe-based BMGs, 316L SS, and TC4 in Hank's solution at 37 °C. These tests, conducted in a thermostatic water bath under an atmospheric environment, provide valuable insights into the electrochemical corrosion parameters, including self-corrosion potential (E_{corr}), self-corrosion current density (I_{corr}), pitting potential (E_{pit}), and corrosion rate (CR), as summarized in Table 2. The I_{corr} values serve as indicators of the corrosion resistance of the materials, while the E_{pit} value and the width of the passivation potential region ($\Delta E = E_{pit} - E_{corr}$) represent the stability of the passive film. The 316L SS shows the highest I_{corr} and the poorest passivation behavior, reflecting its inferior corrosion resistance. In contrast, TC4 demonstrates good corrosion resistance, with I_{corr} values comparable to Ag2 and Ag3. However, TC4 falls short in passivation stability as compared with Ag3. Among the Fe-based BMGs, the incorporation of Ag significantly improves their corrosion performance. As the Ag content increases, I_{corr} decreases, and the corrosion resistance improves. Ag3 displays the lowest I_{corr} and the most stable passivation behavior, indicating superior corrosion resistance. Its wide passivation potential range and minimal I_{corr} highlight its effectiveness in resisting corrosive environments. These findings underscore the potential of Ag3 for biomedical applications, as it outperforms both 316L SS and TC4 in corrosion resistance and passivation stability.

Table 2. Electrochemical parameters obtained from the potentiodynamic polarization curves in Hank's solution at 37 °C. (I_{corr}: corrosion current density; E_{corr}: corrosion potential; E_{pit}: pitting potential; CR: corrosion rate).

Alloys	I_{corr} ($\times 10^{-6}$ A/cm^2)	E_{corr} (mV)	E_{pit} (mV)	$\Delta E = E_{pit} - E_{corr}$ (mV)	CR (10^{-3} mm/a)
316L SS	4.09 ± 0.16	−229 ± 21	518 ± 13	747 ± 35	4.56 ± 0.2
TC4	0.81 ± 0.22	−76 ± 13	-	-	0.90 ± 0.3

Table 2. *Cont.*

Alloys	I_{corr} ($\times 10^{-6}$ A/cm^2)	E_{corr} (mV)	E_{pit} (mV)	$\Delta E = E_{pit} - E_{corr}$ (mV)	CR (10^{-3} mm/a)
Ag0	3.43 ± 0.17	-115 ± 14	1062 ± 9	1177 ± 22	3.77 ± 0.4
Ag1	1.23 ± 0.18	-189 ± 14	1004 ± 12	1193 ± 25	1.34 ± 0.2
Ag2	0.62 ± 0.13	-81 ± 8	571 ± 11	652 ± 20	0.69 ± 0.4
Ag3	0.51 ± 0.11	-166 ± 12	473 ± 15	639 ± 25	0.56 ± 0.2

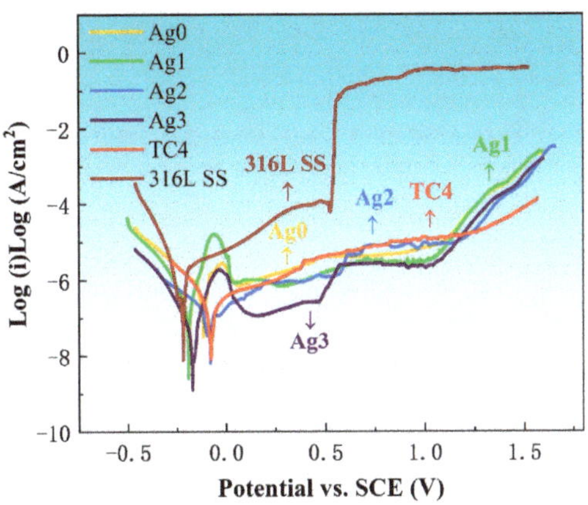

Figure 4. Potentiodynamic polarization curves of the Fe-based BMGs, 316L SS, and TC 4 in Hank's solution at 37 °C.

2.3. Metal Ion Release

Figure 5 illustrates the ion concentrations released into Hank's solution from the Fe-based BMGs, 316L SS, and TC4 after potentiodynamic polarization. The results reveal that 316L SS releases significantly higher concentrations of Fe, Cr, Mo, Mn, and Ni ions as compared with the Fe-based BMGs. This elevated ion release is attributed to the inferior corrosion resistance of 316L SS, which leads to a greater dissolution of metal ions in physiological environments. Notably, the release of Ni ions is particularly concerning, as Ni can interfere with enzymatic activity and trigger inflammatory and allergic responses [23]. Consequently, the high levels of Ni released from 316L SS significantly compromises its biocompatibility. In contrast, both TC4 and Fe-based BMGs demonstrate minimal ion release, a direct result of their superior corrosion resistance. Among the Fe-based BMGs, those with higher Ag contents exhibit a slightly increased release of Ag ions, reflecting the incorporation of Ag as an antimicrobial agent. Nevertheless, the overall ion release from Fe-based BMGs remains exceptionally low, effectively minimizing the risk of harmful ions leaching into the body. These findings highlight the potential of the Fe-based BMGs as safe and reliable materials for biomedical implants, offering a combination of excellent corrosion resistance and minimal ion release, which are critical for maintaining biocompatibility and preventing adverse reactions in clinical applications.

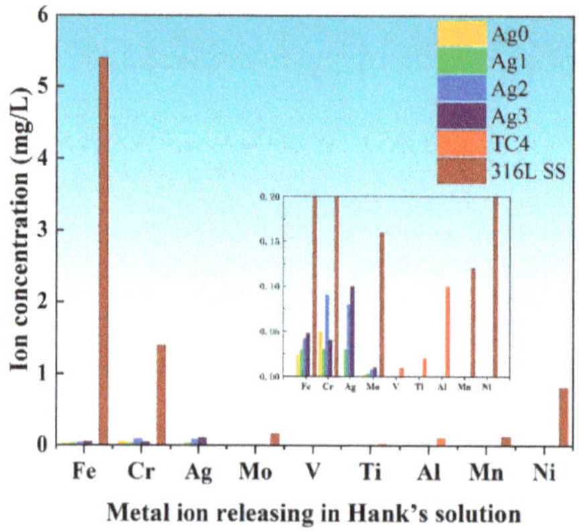

Figure 5. The concentrations of ions dissolved into the Hank's solution after potentiodynamic polarization. The insets are the enlarged-scale views to clearly show the low ion concentrations.

2.4. Biocompatibility

Figure 6 illustrates the cell viability ratio (R_{cv}) of the Fe-based BMGs, 316L SS, and TC4 extracts over 1, 3, 5, and 7 days of incubation. The survival rate for all experimental groups ranged from 75% to 99%, corresponding to a toxicity grade of 0 or I, which aligns with the national standards for toxicity testing of medical and hygienic materials [24]. Among the tested materials, 316L SS consistently exhibited the lowest cell viability as compared to the control, attributed to the release of cytotoxic ions such as Ni. In contrast, the cell viability of the Fe-based BMG was 8% and 1% higher than that of 316L SS and TC4, respectively, although a slight decrease was observed with the increasing Ag content. Compared with Ag0, the cell viability of Ag3 decreased by 10%.

Figure 7 presents the adherent morphology of L929 cells under an inverted fluorescence microscope after 1, 3, 5, and 7 days of culture. By day 7, all samples showed a substantial increase in cell density, with cells spreading and extending in a shuttle-like pattern at low density. The results reveal no significant differences between the control and experimental groups, indicating that none of the tested materials had adverse effects on cell growth.

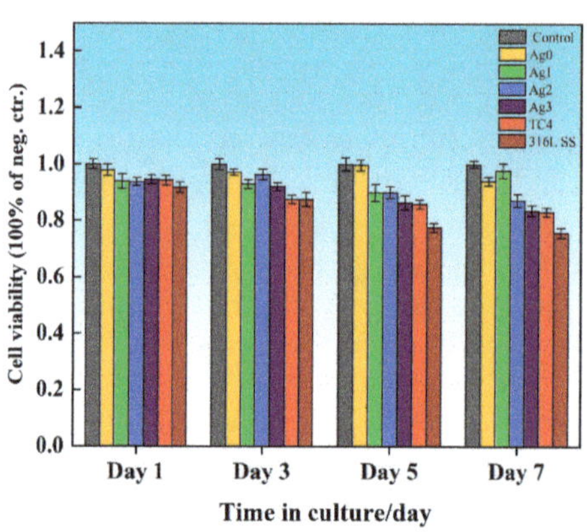

Figure 6. Cytotoxicity of L929 cells in the Fe-based BMGs, 316L SS, and TC4 extraction media for 1, 3, 5, and 7 days.

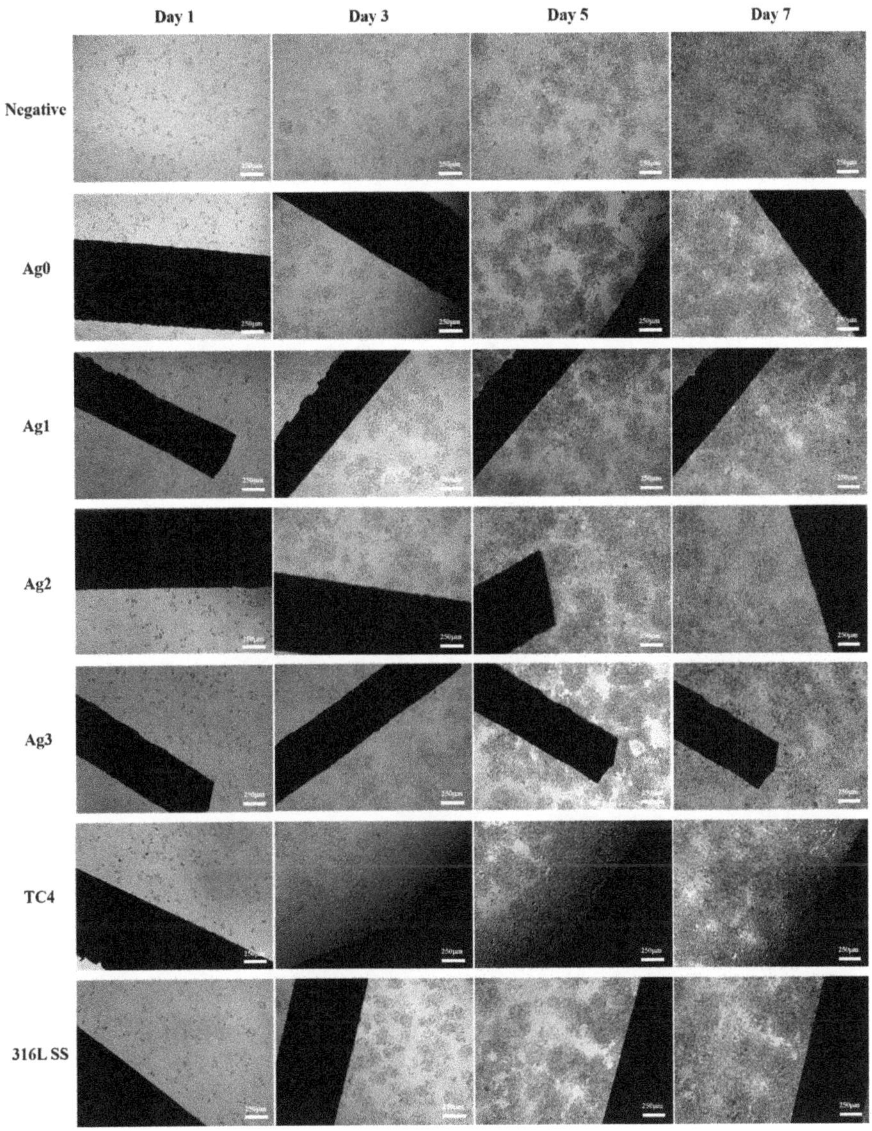

Figure 7. Morphologies of L929 cells cultured with the Fe-based BMGs, 316L SS, and TC4 samples' extractions after 1, 3, 5, and 7 days.

To further examine cell attachment and morphology, Figure 8 provides SEM images of L929 cells adhering to the surfaces of the Fe-based BMGs, TC4, and 316L SS after 7 days of culture. The surface of the Fe-based BMGs was nearly completely covered with cells, reflecting their superior biocompatibility. In contrast, the surfaces of TC4 and 316L SS showed fewer adherent cells. The reduced cell attachment on TC4 is attributed to its biological inertness, which requires surface activation to improve cell adhesion. Meanwhile, 316L SS significantly reduced cell viability and attachment due to the release of toxic ions, such as Ni and Cr. SEM observations also showed intact cell morphology on all Fe-based BMG surfaces, indicating healthy cell growth without signs of rupture.

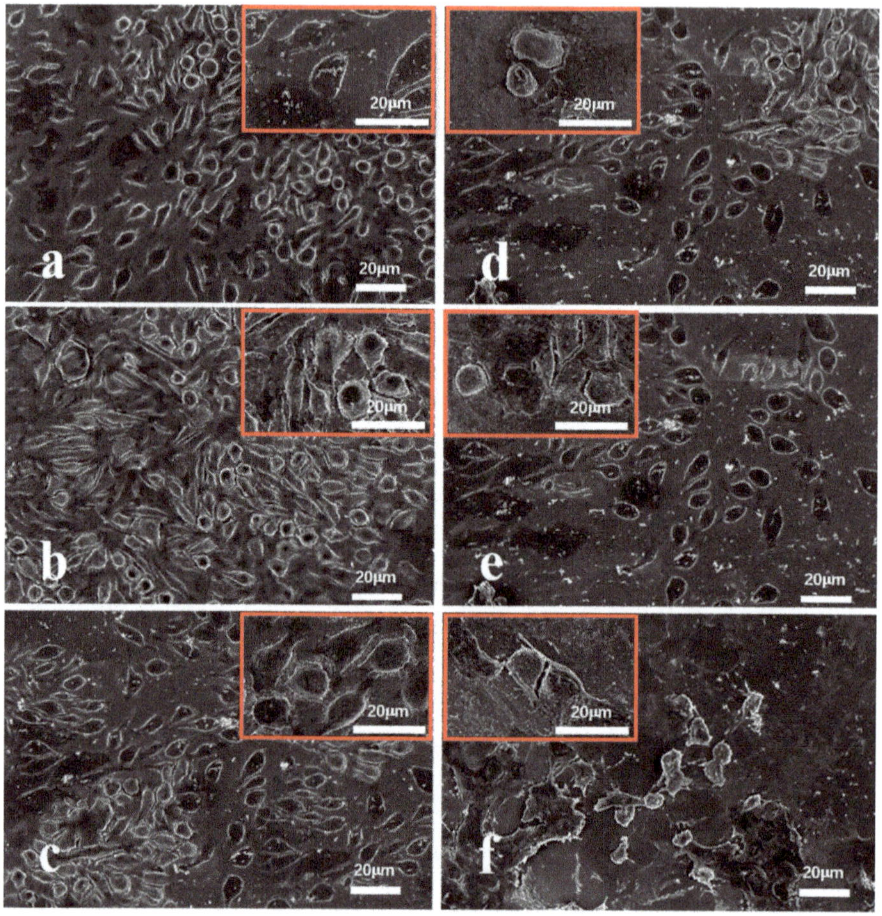

Figure 8. SEM images of L929 cells adhesion on the surfaces of (**a**) Ag0, (**b**) Ag1, (**c**) Ag2, (**d**) Ag3, (**e**) TC4, and (**f**) 316L SS and their corresponding local enlarged images.

These findings collectively highlight the superior biocompatibility of the Fe-based BMGs as compared with TC4 and 316L SS, demonstrated by higher cell attachment, better cell morphology, and stronger cell activity across all tests.

2.5. Antibacterial Activity

Figure 9 illustrates the bacterial inhibition rates of the Fe-based BMGs, 316L SS, and TC4 against *E. coli* and *S. aureus* at 3, 12, and 24 h of incubation. All six materials exhibit antibacterial properties to varying degrees. The Fe-based BMGs reduce bacterial attachment due to their amorphous microstructure, which limits the growth and proliferation of bacteria. Similarly, TC4 impedes bacterial attachment, owing to its inert surface. Meanwhile, 316L SS exhibits antibacterial performance partly due to the release of ions such as Cr, Fe, Mo, and Ni. However, its effectiveness remains inferior to the Ag-containing Fe-based BMGs.

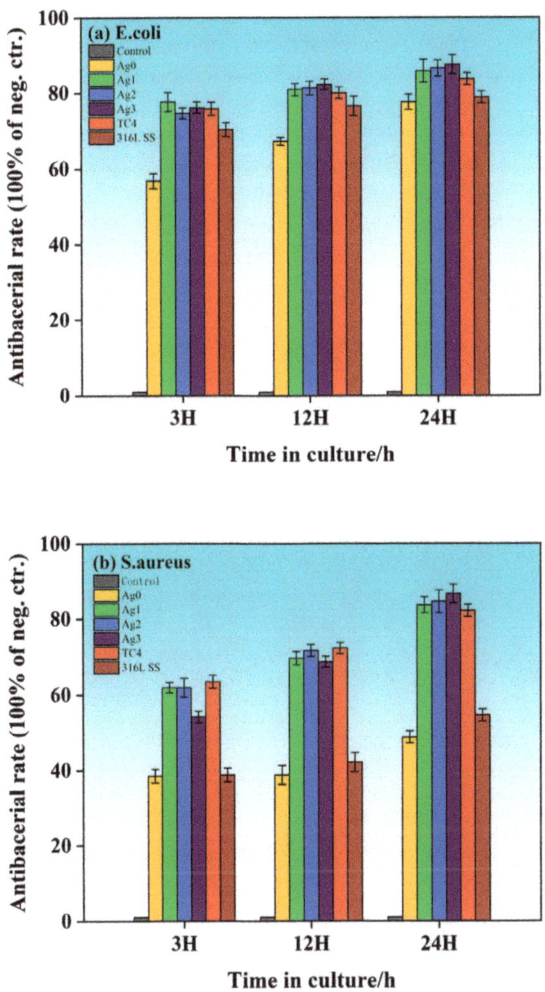

Figure 9. Antibacterial rates of the Fe-based BMGs, 316L SS, and TC4: (**a**) *E. coli* and (**b**) *S. aureus*.

The addition of Ag significantly enhances the antimicrobial performance of the Fe-based BMGs. At the early stage (3 h), the bacterial inhibition rate is similar across the samples, except for the Ag0, likely due to insufficient release of Ag ions. Over time (12–24 h), the antimicrobial performance of the Ag-containing Fe-based BMGs improves markedly, surpassing both TC4 and 316L SS. This enhanced performance can be attributed to the release of Ag^+ ions, which trigger the reactive oxygen species (ROS) mechanism. ROS, including superoxide radicals, hydrogen peroxide, and hydroxyl radicals, disrupt bacterial membranes and induce oxidative damage, ultimately leading to cell death. Among the samples, the Ag3 demonstrates the highest antibacterial effect, achieving an inhibition rate of 87.5% against *E. coli*. Similarly, for *S. aureus*, the Ag3 exhibits an antimicrobial rate of 86.7%, significantly outperforming 316L SS (54.6%) and the Ag0 (48.7%).

To further analyze the antibacterial behavior, bacterial adhesion was observed using FESEM after 24 h of incubation. Figure 10 shows minimal *E. coli* adhesion on the surfaces of the Ag-containing Fe-based BMGs, with only a few bacterial colonies observed on the Ag3. In contrast, the Ag0 exhibits higher bacterial aggregation, reflecting its weaker antimicrobial performance. Figure 11 highlights similar trends for *S. aureus*. The Ag0 and 316L SS show substantial bacterial adhesion, with dense colonies covering their surfaces. Interestingly, the diameter of *S. aureus* colonies on the Ag0 is relatively small (0.2 μm), indicating limited bacterial growth. However, as the Ag content increases, bacterial adhesion is significantly reduced. The surface of the Ag3 shows only a few isolated bacterial colonies, demonstrating its superior antimicrobial effectiveness.

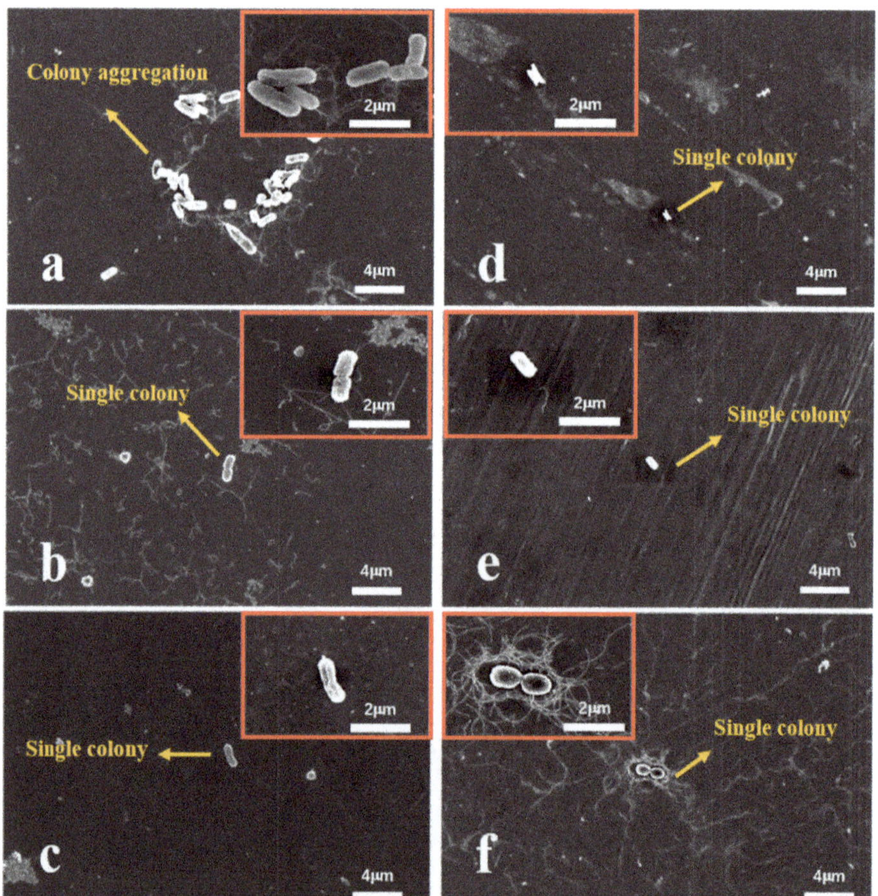

Figure 10. SEM images of *E. coli* adhesion on the surfaces of (**a**) Ag0, (**b**) Ag1, (**c**) Ag2, (**d**) Ag3, (**e**) TC4, and (**f**) 316L SS and their corresponding local enlarged images.

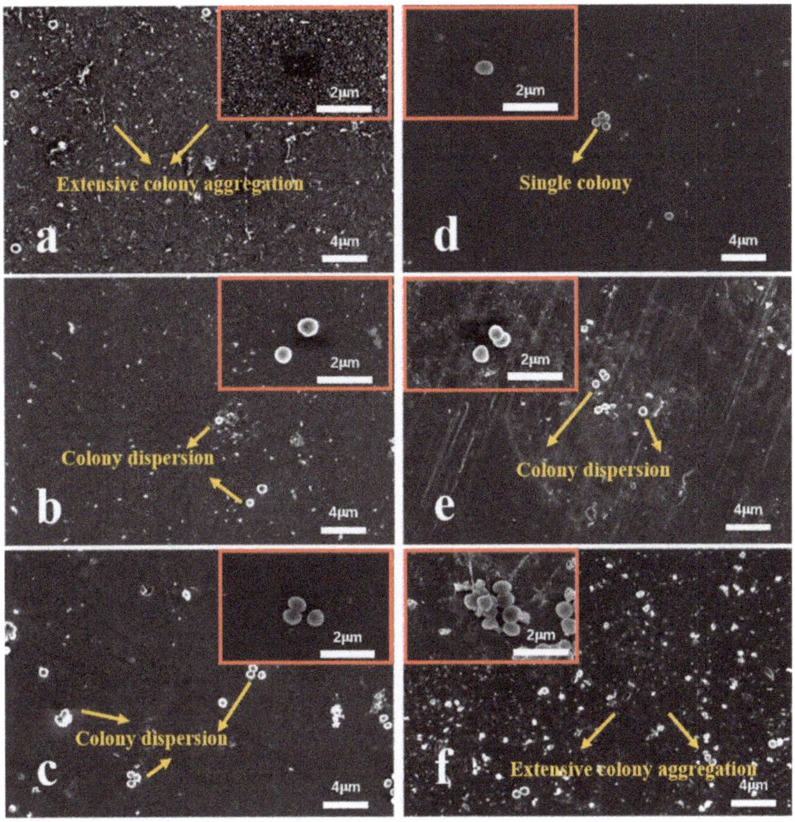

Figure 11. SEM images of *S. aureus* adhesion on the surfaces of (**a**) Ag0, (**b**)Ag1, (**c**) Ag2, (**d**) Ag3, (**e**) TC4, and (**f**) 316L SS and their corresponding local enlarged images.

3. Materials and Methods

3.1. Material Preparation

$Fe_{55-x}Cr_{20}Mo_5P_{13}C_7Ag_x$ (x = 0, 1, 2, 3 at.%) master alloy ingots were synthesized through induction melting of high-purity raw materials, including Fe (99.9 wt.%), Cr (99.9 wt.%), Mo (99.9 wt.%), graphite (99.9 wt.%), Ag (99.9 wt.%), and Fe_3P (99.5 wt.%), under a high-purity argon atmosphere after a vacuum of approximately 10^{-5} Pa. The resulting alloy ingots were subsequently processed into glassy alloy rods of varying diameters using a copper-mold suction casting technique under the protection of pure argon gas.

3.2. Material Characterization

The amorphous structure of the as-cast alloy samples was verified using X-Ray diffraction (XRD) analysis performed on a Bruker D8 Advance diffractometer (Bruker AXS GmbH, Karlsruhe, Germany) equipped with Cu K_α radiation (operating at 30 kV and 30 mA) at room temperature. The thermal behavior of the samples was analyzed via differential scanning calorimetry (DSC, NETZSCH, Görlitz, Germany) under an argon atmosphere, with a heating rate of 0.33 K/s.

To evaluate the mechanical properties, compression testing was conducted. The test samples, each with a diameter of 1 mm, were precisely cut to lengths of 2.5 mm. The ends of these samples were meticulously polished and smoothened using 2000-grit sandpaper to achieve a final length of 2 mm. Compression tests were performed at room temperature using an electronic universal testing machine (ETM105D, Wance, Shenzhen, China). To ensure data reliability, each test was repeated five times for each alloy composition.

3.3. Electrochemical Measurements

Electrochemical testing was conducted to evaluate the corrosion resistance of the Fe-based BMGs, 316L SS, and TC4, all prepared as 1 mm diameter bars. The experiments were performed using a Zahner Zennium X (Zahner, Neuss, Germany) electrochemical workstation in a three-electrode cell configuration. The setup included a platinum electrode as the auxiliary electrode, a saturated calomel electrode as the reference, and the test sample as the working electrode. The temperature was maintained at 37 °C using a water bath, with Hank's solution (simulated body fluids, pH 7.4, supplied by China Beijing Biotech Ltd., Beijing, China) serving as the electrolyte.

Before testing, all samples were thoroughly polished with progressively finer grit sandpaper (600, 800, 1200, 2000, 3000, 4000, and 6500), followed by cleaning with alcohol via brief ultrasonication and air drying to ensure clean, smooth surfaces. The samples were immersed in the electrolyte for 30 min to establish a stable open-circuit potential before electrochemical polarization. The polarization curves were recorded at a scan rate of 1 mV/s over a potential range of −0.5 V to 1.6 V. The electrochemical corrosion parameters, including corrosion potential (E_{corr}), corrosion current density (I_{corr}), pitting potential (E_{pit}), and corrosion rate (CR), were derived from the kinetic potential polarization curves to comprehensively assess the corrosion resistance of the tested materials.

3.4. ICP Test

The concentration of ions released into Hank's solution following kinetic potential polarization was quantified using full-spectrum, direct-reading inductively coupled plasma atomic-emission spectrometry (ICP-AES) with a PerkinElmer Optima 8000 spectrometer (PerkinElmer, Waltham, MA, USA). Each experimental group was measured in triplicate, and the results were averaged to ensure accuracy and reliability.

3.5. Cytocompatibility Evaluation

3.5.1. Cell Culture

Mouse embryonic fibroblasts (L929 cell line, Servicebio) were maintained in a basal medium consisting of Minimum Essential Medium (MEM, Servicebio) supplemented with 10% fetal bovine serum (FBS, Servicebio) and 1% penicillin–streptomycin antibiotics (Servicebio).The mouse embryonic fibroblasts (L929 cell line) were obtained from Servicebio (Wuhan, China)(Cat#: STCC20025P). The NCTC clone 929 [L cell, L-929, derived from C3H/An L species] cell line is a clone of the L cell line created in March of 1948. The L cell line is among the first established continuous cell lines, with L-929 being one of the early clonal derivatives. The parent L cell line was developed from the subcutaneous loose connective and adipose tissue of a 100-day-old male C3H/An mouse. At passage 95, the L cell line was employed to isolate single cells by capillary cloning to generate the L-929 clone. The cells were cultured in a humidified incubator at 37 °C with 5% CO_2 to ensure optimal growth conditions. After initial incubation, the cells were washed with phosphate-buffered saline (PBS, Servicebio) and replenished with fresh medium. Following approximately 48 h of incubation, the cells were allowed to proliferate until they reached confluence at the bottom of the culture flask.

To harvest the cells, 0.25% trypsin solution was used to enzymatically detach them from the culture surface. The reaction was terminated by adding a growth medium comprising 44 mL MEM, 5 mL FBS, and 1 mL penicillin–streptomycin solution. The cells were subsequently counted using the plate counting method, and the cell concentration was adjusted to 10,000 cells per well for downstream experiments.

3.5.2. Cell Counting Kit-8 (CCK-8) Test

To ensure consistent and controlled experimental conditions, the samples used in this study, including the Fe-based BMGs, 316L SS, and TC4, were prepared as 3 mm × 1 mm sheets and polished to a mirror finish. The samples were thoroughly cleaned using an ultrasonic machine with anhydrous ethanol and deionized water for 10 min, followed by drying and sterilization in an autoclave to guarantee sterility. Three replicates of each sample were prepared and handled in an ultra-clean biological laboratory to maintain contamination-free conditions.

The prepared samples were inoculated into 24-well plates along with a blank group (no cells) and a control group (cells without material). Mouse embryonic fibroblast cells (5000 cells/well) were seeded into the 24-well plates and incubated in a humidified incubator at 37 °C with 5% CO_2. Cell viability was assessed at four time points: 1, 3, 5, and 7 days. At each time point, 90 μL of cell suspension from each well was transferred to a 96-well plate, and 10 μL of CCK-8 reagent was added. The mixture was homogenized and incubated at 37 °C for 2 h. Following incubation, spectrophotometric absorbance measurements were taken at 450 nm using an enzyme-linked microplate reader.

For each experiment, three replicates (sample and control) were analyzed to ensure data reliability. The relative cell viability (R_{cv}) of each sample was calculated using the following formula: R_{cv} = (cell activity in experimental extract)/(cell activity in negative control). Calculate the standard deviation of the data from multiple experimental results using the standard deviation formula, and then use the standard deviation to represent the error. In the graph, set the length of the error bars to the standard deviation.

3.5.3. Cell Adhesion

Cell growth and adhesion were systematically evaluated at intervals of 1, 3, 5, and 7 days during the CCK-8 experiments. Observations and photographic documentation were performed using inverted fluorescence microscopy (Leica DMi8, Wetzlar, Germany). Following the 7-day incubation period, the samples co-cultured with cells were carefully removed and rinsed three times with phosphate-buffered saline (PBS) to eliminate any non-adherent cells. Subsequently, the adherent cells on the sample surfaces were fixed with 3% glutaraldehyde at room temperature for 1.5 h, protected from light to prevent degradation.

To prepare the samples for further analysis, they were sequentially dehydrated using ethanol solutions of increasing concentrations (30%, 50%, 70%, 95%, and 100%), with each step lasting 15 min at room temperature. Once dehydrated, the samples were dried in a vacuum atmosphere to preserve their structures. The final prepared samples were examined using an ultra-high-resolution cold field emission scanning electron microscope (FESEM, Hitachi, Tokyo, Japan) to investigate the surface morphology and growth patterns of the adhering cells.

3.6. Antibacterial Evaluation

3.6.1. Bacterial Culture

Healthcare-associated infections are frequently caused by pathogenic microorganisms, with *E. coli* and *S. aureus* being among the most prevalent culprits. To investigate the antimicrobial properties of the developed materials, this study employed *E. coli* (ATCC) and *S. aureus* (ATCC) as representative experimental bacteria, encompassing both Gram-negative and Gram-positive strains.

Bacterial cultures were grown on Luria–Bertani (LB) agar medium plates (Servicebio). The LB solid medium was prepared by dissolving the dry powder in purified water, followed by sterilization in an autoclave at 121 °C for 30 min. After cooling the sterilized medium to 50–60 °C, it was poured into sterile Petri dishes under ultra-clean conditions and allowed to solidify. Bacteria were extracted from frozen stock cultures and inoculated onto the prepared LB plates under aseptic conditions. The inoculated plates were incubated at 37 °C for 24 h in a constant-temperature incubator, facilitating the formation of distinct bacterial colonies.

On the subsequent day, fresh broth cultures were prepared to ensure viable bacterial populations for the experiments. This was achieved by diluting bacterial suspensions in fresh LB broth to a final concentration of 1×10^5 CFU/mL, corresponding to a spectrophotometric optical density (OD) of 0.001 at 600 nm. This standardized preparation ensured reproducibility and reliability in evaluating the antimicrobial efficacy of the tested materials.

3.6.2. Antibacterial Activity Evaluation

To control the effective area, the samples (Fe-based BMGs, 316L SS, and TC4) were prepared as 3 mm × 1 mm sheets and polished to a mirror finish. All samples were cleaned with anhydrous ethanol and deionized water in an ultrasonic cleaner for 10 min, dried, and sterilized in an autoclave to ensure sterility. The sterilized samples were submerged in 24-well plates containing 1 mL of *E. coli* or *S. aureus* LB medium with a bacterial concentration of 1×10^5 CFU/mL. Blank groups (no bacteria) and control groups (bacteria without material) were also included. The plates were incubated statically at 37 °C for 24 h. For each experiment, three replicates (sample and control) were analyzed to ensure data reliability. The following formula was employed: Calculate the standard deviation of the data from multiple experimental results using the standard deviation formula, and then use the standard deviation to represent the error. In the graph, set the length of the error bars to the standard deviation.

Bacterial inhibition was monitored spectrophotometrically at 600 nm, with measurements taken after 3 h (early stage), 12 h (intermediate stage), and 24 h (late stage). The inhibition rate was calculated using the following formula:

$$\text{inhibition rate}(\%) = \frac{\text{OD value of experimental group} - \text{OD value of blank group}}{\text{OD value of control group} - \text{OD value of blank group}} \times 100\%$$

After 24 h, the samples were gently rinsed three times with PBS, and surface-adhering bacteria were fixed with 3% glutaraldehyde at room temperature in the dark for 1.5 h. The samples were sequentially dehydrated with ethanol solutions (30%, 50%, 70%, 95%, and 100%) for 15 min each at room temperature and dried under vacuum. Bacterial adhesion and growth were subsequently analyzed using an ultra-high-resolution cold field emission scanning electron microscope (FESEM, Hitachi, Tokyo, Japan).

4. Conclusions

In this study, $Fe_{55-x}Cr_{20}Mo_5P_{13}C_7Ag_x$ (x = 0, 1, 2, 3 at.%, denoted as Ag0, Ag1, Ag2, and Ag3, respectively) bulk metallic glasses (BMGs) were prepared to investigate the effects of Ag addition on their antimicrobial and biomedical properties. The key conclusions are as follows:

(1) The glass-forming ability (GFA) of the Fe-based BMGs decreased with increasing Ag content, as indicated by a reduction in the critical diameter for glass formation from

2.0 mm for Ag0 to 1.0 mm for Ag3. This decline was attributed to the positive enthalpy of the mixing between Ag and Fe, which promotes crystalline phase precipitation.

(2) The Fe-based BMGs exhibited superior corrosion resistance as compared with 316L SS and TC4 in Hank's solution at 37 °C. The Ag3 demonstrated the lowest corrosion current density and the most stable passivation behavior, making it highly resistant to localized corrosion.

(3) The Fe-based BMGs showed excellent biocompatibility, outperforming 316L SS and TC4 in terms of cell viability and adhesion. SEM analysis confirmed superior cell attachment and morphology on the Fe-based BMG surfaces, and minimal harmful ion release ensured their safety as implant materials.

(4) The addition of Ag significantly enhanced the antimicrobial properties of the Fe-based BMGs. Ag3 achieved the highest bacterial inhibition rates, with 87.5% against *E. coli* and 86.7% against *S. aureus*, surpassing both 316L SS and TC4. This was attributed to the Ag^{+} ions triggering the production of reactive oxygen species (ROS), which disrupt bacterial membranes and induce cell death.

(5) Combining exceptional mechanical properties, corrosion resistance, biocompatibility, and antimicrobial performance, the present Ag-containing Fe-based BMGs, particularly Ag3, emerge as promising candidates for next-generation biomedical implants. These materials provide a cost-effective, high-performance alternative to conventional implants, addressing critical challenges such as bacterial infections and implant failures.

Additionally, there are several promising research directions worth further exploration in this study, which are summarized as follows:

(1) As a potential orthopedic implant material, its osteogenic ability can be further assessed by culturing osteoblasts and testing their cellular activity in the samples.

(2) Biofilms play a crucial role in antibacterial research. Therefore, future studies could focus on dynamic biofilm research and design multifunctional antibacterial biofilms on the material surface.

(3) This study lacks in vivo experimental validation, which limits a comprehensive understanding of the material's performance in a real biological environment. Future research should strengthen in vivo experiments to further validate the clinical relevance of the in vitro results, providing a more solid foundation for the application of Fe-based amorphous alloys in the biomedical field.

Author Contributions: Conceptualization, L.J.; methodology, L.J.; formal analysis, L.J.; data curation, L.J., X.F. and X.L.; writing—original draft preparation, L.J.; writing—review and editing, Q.L.; supervision, T.J. and Q.W.; funding acquisition, Q.L. and T.J. All authors have read and agreed to the published version of the manuscript.

Funding: This research was supported by the National Natural Science Foundation of China (Grant No. 52261033), the Special Fund for Innovation Environment Construction (Talent and Base) in the Xinjiang Uyghur Autonomous Region—Construction of Science and Technology Innovation Bases (Resource Sharing Platform Development) (Grant No. PT2304), and the Open Topics Project of the Xinjiang Medical University Institute of Medical Sciences (Grant No. YXYJ20230202).

Institutional Review Board Statement: Not applicable.

Informed Consent Statement: Not applicable.

Data Availability Statement: Data is contained within the article.

Conflicts of Interest: The authors declare no conflicts of interest.

References

1. Grosgogeat, B.; Reclaru, L.; Lissac, M.; Dalard, F. Measurement and evaluation of galvanic corrosion between titanium/Ti6Al4V implants and dental alloys by electrochemical techniques and auger spectrometry. *Biomaterials* **1999**, *20*, 931–941.
2. Long, M.; Rack, H.J. Titanium alloys in total joint replacement—A materials science perspective. *Biomaterials* **1998**, *19*, 1621–1639. [PubMed]
3. Li, S.; Wei, Q.; Li, Q.; Jiang, B.; Chen, Y.; Sun, Y. Development of Fe-based bulk metallic glasses as potential biomaterials. *Mater. Sci. Eng. C* **2015**, *52*, 235–241. [CrossRef]
4. Gottenbos, B.; van der Mei, H.C.; Klatter, F.; Grijpma, D.W.; Feijen, J.; Nieuwenhuis, P.; Busscher, H.J. Positively charged biomaterials exert antimicrobial effects on gram-negative bacilli in rats. *Biomaterials* **2003**, *24*, 2707–2710. [CrossRef] [PubMed]
5. Hetrick, E.M.; Schoenfisch, M.H. Reducing implant-related infections: Active release strategies. *Chem. Soc. Rev.* **2006**, *35*, 780–789. [CrossRef]
6. Han, K.; Jiang, H.; Wang, Y.; Qiang, J.; Yu, C. Antimicrobial Zr-based bulk metallic glasses for surgical devices applications. *J. Non-Cryst. Solids* **2021**, *564*, 120827. [CrossRef]
7. Normark, B.H.; Normark, S. Evolution and spread of antibiotic resistance. *J. Intern. Med.* **2002**, *252*, 91–106. [CrossRef]
8. Li, H.F.; Zheng, Y.F. Recent advances in bulk metallic glasses for biomedical applications. *Acta Biomater.* **2016**, *36*, 1–20. [CrossRef]
9. Sumner, D.R. Long-term implant fixation and stress-shielding in total hip replacement. *J. Biomech.* **2015**, *48*, 797–800. [CrossRef]
10. Chiang, P.T.; Chen, G.J.; Jian, S.R.; Shih, Y.H.; Jang JS, C.; Lai, C.H. Surface Antimicrobial Effects of $Zr_{61}Al_{7.5}Ni_{10}Cu_{17.5}Si_4$ Thin Film Metallic Glasses on *Escherichia coli*, *Staphylococcus aureus*, *Pseudomonas aeruginosa*, *Acinetobacter baumannii* and *Candida albicans*. *Fooyin J. Health Sci.* **2010**, *2*, 12–20.
11. Fang, H.; Hui, X.; Chen, G. Effects of Mn addition on the magnetic property and corrosion resistance of bulk amorphous steels. *J. Alloys Compd.* **2008**, *464*, 292–295. [CrossRef]
12. Gorejová, R.; Haverová, L.; Oriňaková, R.; Oriňak, A.; Oriňak, M. Recent advancements in Fe-based biodegradable materials for bone repair. *J. Mater. Sci.* **2018**, *54*, 1913–1947. [CrossRef]
13. Jiang, J.; Zhang, C.; Zeng, G.-M.; Gong, J.-L.; Chang, Y.-N.; Song, B.; Deng, C.-H.; Liu, H.-Y. The disinfection performance and mechanisms of Ag/lysozyme nanoparticles supported with montmorillonite clay. *J. Hazard. Mater.* **2016**, *317*, 416–429. [CrossRef] [PubMed]
14. Berger, T.J.; Spadaro, J.A.; Bierman, R.; Chapin, S.E.; Becker, R.O. Antifungal Properties of Electrically Generated Metallic Ions. *Antimicrob. Agents Chemother.* **1976**, *10*, 856–860. [PubMed]
15. Zhao, G.; Stevens, S.E., Jr. Multiple parameters for the comprehensive evaluation of the susceptibility of *Escherichia coli* to the silver ion. *BioMetals* **1997**, *11*, 27–32.
16. Zhao, Y.; Bunch, T.D.; Isom, S.C. Effects of electrical biostimulation and silver ions on porcine fibroblast cells. *PLoS ONE* **2021**, *16*, e0246847. [CrossRef]
17. Akhavan, O. Lasting antibacterial activities of Ag–TiO2/Ag/a-TiO2 nanocomposite thin film photocatalysts under solar light irradiation. *J. Colloid Interface Sci.* **2009**, *336*, 117–124. [CrossRef]
18. Chen, J.; Wan, G.; Leng, Y.; Yang, P.; Sun, H.; Wang, J.; Huang, N.; Chen, H.; Tang, R. Antithrombogenic investigation and biological behavior of cultured human umbilical vein endothelial cells on Ti-O film. *Sci. China Ser. E* **2006**, *49*, 20–28. [CrossRef]
19. Jones, S.A.; Bowler, P.G.; Walker, M.; Parsons, D. Controlling wound bioburden with a novel silver-containing Hydrofiber® dressing. *Wound Repair Regen.* **2004**, *12*, 288–294. [CrossRef]
20. Takeuchi, A.; Inoue, A. Classification of Bulk Metallic Glasses by Atomic Size Difference, Heat of Mixing and Period of Constituent Elements and Its Application to Characterization of the Main Alloying Element. *Mater. Trans.* **2005**, *46*, 2817–2829.
21. Pradeep, K.G.; Herzer, G.; Choi, P.; Raabe, D. Atom probe tomography study of ultrahigh nanocrystallization rates in FeSiNbBCu soft magnetic amorphous alloys on rapid annealing. *Acta Mater.* **2014**, *68*, 295–309. [CrossRef]
22. Cheng, Y.Q.; Ma, E. Atomic-level structure and structure–property relationship in metallic glasses. *Prog. Mater. Sci.* **2011**, *56*, 379–473. [CrossRef]
23. Wang, S.L.; Li, H.X.; Zhang, X.F.; Yi, S. Effects of Cr contents in Fe-based bulk metallic glasses on the glass forming ability and the corrosion resistance. *Mater. Chem. Phys.* **2009**, *113*, 878–883. [CrossRef]
24. Sharma, A.; Luthra, G. Significance of ISO 10993 Standards in Ensuring Biocompatibility of Medical Devices: A Review. *J. Pharm. Res. Int.* **2023**, *35*, 23–34. [CrossRef]

inorganics MDPI

Article

Luminescence Study of Hydrogenated Silicon Oxycarbide (SiO_xC_y:H) Thin Films Deposited by Hot Wire Chemical Vapor Deposition as Active Layers in Light Emitting Devices

Juan R. Ramos-Serrano [1], Yasuhiro Matsumoto [2], Alejandro Ávila [2], Gabriel Romero [2], Maricela Meneses [2], Alfredo Morales [1], José A. Luna [3], Javier Flores [4,5], Gustavo M. Minquiz [4,5] and Mario Moreno-Moreno [1,*]

[1] Electronics Coordination, Instituto Nacional de Astrofísica, Óptica y Electrónica, Tonantzintla, Puebla 72840, Mexico; juan.ramos@inaoep.mx (J.R.R.-S.); alfredom@inaoep.mx (A.M.)
[2] Electrical Engeneering Department, Centro de Investigación y de Estudios Avanzados del IPN, Mexico City 07360, Mexico; ymatsumo@cinvestav.mx (Y.M.); aavila@cinvestav.mx (A.Á.); gromero@cinvestav.mx (G.R.); marmem369@gmail.com (M.M.)
[3] Centro de Investigaciones en Dispositivos Semiconductores (CIDS-ICUAP), Benemérita Universidad Autónoma de Puebla (BUAP), Col. San Manuel, Cd. Universitaria, Av. San Claudio y 14 Sur, Edificios IC5 y IC6, Puebla 72570, Mexico; jose.luna@correo.buap.mx
[4] Ingeniería Mecánica, Tecnológico Nacional de México/I.T. Puebla, Av. Tecnológico #420 Col. Maravillas, Puebla 72220, Mexico; javier.flores@puebla.tecnm.mx (J.F.); gminquiz@yahoo.com (G.M.M.)
[5] Área de Ingeniería, Benemérita Universidad Autónoma de Puebla, Ciudad Universitaria, Blvd. Valsequillo y Esquina Av. San Claudio s/n, Col. San Manuel, Puebla 72570, Mexico
* Correspondence: mmoreno@inaoep.mx; Tel.: +52-(222)-266-31-00

Abstract: The obtention of luminescent SiO_xC_y:H thin films deposited by the HW-CVD technique is reported here. We study the effect of different monomethyl-silane (MMS) flow rates on the films properties. An increase in the emission bandwidth and a red-shift was observed when the MMS flow increased. The luminescence was related to optical transitions in band tail states and with less contribution from quantum confinement effects. After, the films were annealed at 750 °C in nitrogen. The annealed film deposited at the highest MMS flow showed an emission spectrum like the as-deposited film, suggesting the same emission mechanisms. By contrast, the annealed film deposited at the lowest MMS flow showed two emission bands. These bands are due to the activation of radiative defects related to oxygen-deficient centers. MOS-like structures were fabricated as electroluminescent devices using the annealed films. Only the structure of the film with the highest carbon content showed light emission in a broad band in the visible spectrum region in forward bias, with a maximum centered close to 850 nm. The light emission mechanism was related to electron thermalization in the band tail states and a direct hole injection into deep states. The trap-assisted tunneling, Poole–Frenkel emissions and Fowler–Nordheim tunneling were proposed as the charge transport mechanism.

Keywords: hot wire chemical vapor deposition; monomethyl-silane; SiOxCy:H; electroluminescence

1. Introduction

Photonics emerges as an alternative to conventional electronics for digital communications, enabling a greater bandwidth, longer distance connections, and lower energy consumption. Specifically, silicon photonics is related to the ability to emit and detect photons using devices fabricated through processes compatible with the existing silicon–CMOS infrastructure. Currently, some silicon photonic devices are marketed; however, for light-emitting devices they continue using III–V compound-semiconductors. Today, there is a growing interest in research for developing light-emitting sources with silicon-based materials, because this allows a better integration in silicon photonics chip manufacturing, taking advantage of the current silicon–CMOS infrastructure. Nowadays, silicon oxycarbide has attracted attention as a functional silicon-based material for its potential application in the development of silicon photonics. This is mainly due to the capability to modify its optical properties by changing the oxygen–carbon ratio. It has been reported that silicon oxycarbide displays an intense photoluminescence in a wide range of the visible spectrum by modifying the above relationship [1]. On the other hand, low-loss optical waveguides have been developed using the capability to adjust the refractive index to values between those obtained for silicon dioxide and silicon carbide (1.5–2.3), by changing the oxygen–carbon ratio [2].

Chemical vapor deposition (CVD) is a widely used technique for a great variety of high-quality thin films, including silicon-based materials such as hydrogenated amorphous silicon (a-Si:H), silicon carbide (SiC), silicon oxide (SiO_x), and silicon nitride (SiN_x) [3–5]. There are different CVD configurations according to the energy source to produce the dissociation of gas species and promote reactions. Among these processes are the thermal energy source in conventional furnaces, atmospheric pressure CVD (AP-CVD) and low pressure CVD (LP-CVD); plasma enhanced CVD (PE-CVD) is also widely used, where plasma is the source of energy, along with hot wire CVD (HW-CVD), which is used to chemically decompose the precursors [6].

Specifically, the HW-CVD technique has aroused interest in silicon-based material deposition for its simplicity, since it has been reported to obtain nano and microcrystalline phases at low substrate deposition temperatures, as low as 200 °C [7,8]. This work studies the structural and optical properties of hydrogenated silicon oxycarbide (SiO_xC_y:H) thin films. These films were deposited using the HW-CVD technique with a monomethyl-silane (MMS), hydrogen and oxygen mixture, varying only the MMS flow rate. For the electro-optical analysis, MOS-like structures were fabricated using the annealed SiO_xC_y:H films as an active layer.

2. Results and Discussion

Hydrogenated silicon oxycarbide (SiO_xC_y:H) thin films were deposited by the HW-CVD technique using monomethyl-silane (MMS) as a single source of silicon and carbon atoms. The MMS flow rate was varied at 3 and 7 sccm. The films were labeled as S3 and S7, as a reference to the MMS flow. In addition, the films were thermally annealed at 750 °C in nitrogen ($S3_{TA}$, $S7_{TA}$). After the characterization of the films, MOS-like structures were fabricated as electroluminescent devices using the annealed films as active layers because these displayed more intense light emissions than as-deposited films. The deposition parameters, thermal annealing, and MOS-like structure fabrication process are described in Section 3.

2.1. SiO_xC_y:H Thin Films Characterization

X-ray photoelectron spectroscopy (XPS) characterization was carried out to determine the elemental composition and binding configuration. Table 1 shows the elemental composition obtained by XPS for the as-deposited and annealed films. A direct relation between the MMS flow rate and the silicon and carbon content was observed for the as-deposited films. This is due to a greater amount of these within the chamber during the depositions. Rates of tantalum coming from the wire were detected for the S3 film. This is because, at a low MMS flow, the tantalum wire showed quick oxidation related to a low oxygen dilution, where the wire showed a rough surface and was dark gray in color. It is possible to use metallic oxides as solid sources in the HW-CVD technique, in this case Ta_2O_5. This process consists in the generation of atomic hydrogen by the dissociation of molecular hydrogen on the hot wire surface. The atomic hydrogen is highly reactive, inducing first a reduction process and, subsequently, the generation of metal hydrides as gaseous precursors [9]. On the other hand, under the deposition conditions of the S7 film, a protective layer of SiC forms on the wire, showing a smooth surface with a golden color, avoiding the incorporation of tantalum into the films. For the annealed films, a slight reduction in the carbon content is observed while the oxygen content increases. This may be due to a substitution of the desorbed elemental carbon and CH_n radicals by oxygen atoms.

Table 1. Elemental composition by XPS of the as-deposited and annealed SiOxCy:H films.

Film	Elemental Composition (% atm.)			
	Si	O	C	Ta
S3	39	47	13	1
S7	43	34	23	-
$S3_{TA}$	39	49	11	1
$S7_{TA}$	44	36	20	-

Additionally, from the XPS measurements in the Si2p window, the valence states of the silicon Si^{n+} (n = 0–4) can be determined with the binding energy. For SiO_x films, these states are associated with five possible matrix configurations, which can be expressed as Si-(Si_{4-n}-O_n) with n = 0–4 [10]. However, due to the incorporation of carbon, it is possible to have a great variety of compounds with different stoichiometries. Other authors relate these valence states to the formation of Si-(C_{4-n}-O_n) compounds with n = 0–4 [11]. Because of this, it is difficult to relate the valence states to a single compound in the SiO_xC_y:H matrix. However, for practical purposes, we consider the simplest compounds as a reference. In this case, elemental silicon (Si-Si_4), silicon carbide (Si-C_4) and stoichiometric silicon dioxide (Si-O_4) can be related to Si^0, Si^{+1}, and Si^{+4} valence states, respectively. As another reference, the binding energy increases with higher oxygen content. Figure 1a,b show the XPS spectra in the Si2p window of the S3 and S7 films. In both cases, the spectra show broad bands related to the formation of sub-stoichiometric oxide and silicon oxycarbide compounds, where the main peak placed close to 102.3 eV (S^{+3}) indicates the presence of compounds like Si–Si–O_3 and C–Si–O_3. The silicon oxidation states suggest the presence of oxygen-deficient centers (ODCs) in the SiO_xC_y:H matrix. These ODCs may correspond to several defect-related emission centers. In addition, the spectrum of the S7 film shows a broader band, where, unlike the S3 film, the contribution of s^0 and s^{+1} bands are observed.

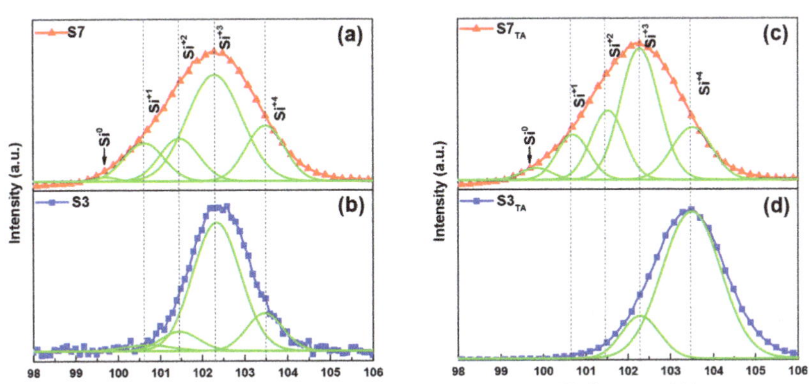

Figure 1. Si2p window in the XPS spectra of the SiO_xC_y:H films (**a**,**b**) as-deposited films and (**c**,**d**) annealed films. The spectra were deconvoluted in components (green lines) related to different silicon valence states.

This indicates that a greater amount of Si–Si and Si–C bonds are formed with increasing MMS flow, which may correspond to the formation of Si/SiC amorphous clusters or nanocrystalline phases. Figure 1c,d show the XPS spectra of the annealed films. The $S3_{TA}$ spectrum shows a shift to a higher binding energy, corresponding to the S^{+4} state. This is due to an increase in the Si–O bonds, where, during the annealing, hydrogen, elemental carbon and different radicals are desorbed and replaced by oxygen atoms. On the other hand, the $S7_{TA}$ film does not show a shift in the maximum of the spectrum; however, the intensity of the contributions related to S^0 and S^{+1} valence states increase with the annealing. In this case, the result suggests an increase in the amount of Si–Si and Si–C bonds as a result of the higher silicon and carbon content.

Fourier-transform infrared (FTIR) spectroscopy measurements were carried out to provide a more detailed structural analysis of the films. Figure 2a shows the FTIR absorbance spectra of the as-deposited films. Due to the amorphous nature of the host matrix and its complex composition, all films display broad and overlapped spectra. The spectra show absorption bands close to 460, 800, and 1080 cm^{-1} related to rocking, bending and stretching (TO) vibrational modes in Si–O–Si bonds [12]. The stretching absorption band shifts to 1052 and 1041 cm^{-1} for S3 and S7 films, respectively. This shift is directly related to the silicon oxide stoichiometry, where values lower than 1080 cm^{-1} (for stoichiometric SiO_2) indicate the presence of many ODCs due to oxygen vacancies and to the change in the bonding characteristics as a result of carbon incorporation [13,14]. Next to this absorption band, a broad shoulder at around 1100 cm^{-1} is commonly attributed to the Si–O–Si stretching (LO) mode [12]. Close to 920 cm^{-1} appears a broad absorption band related to the stretching vibration of silanol (Si–O–H) groups [15]. The spectra show a lower intensity in this band for the S7 film; this is related to a decrease in the amount of Si–O–H bonds when the MMS flow increases, where a higher carbon content favors the formation of Si–C–H bonds. The absorption band corresponding to the Si–C bonds at 800 cm^{-1} is observed for the S7 film [16]. This is in accordance with the XPS measurements. Other absorption bands related to the Si–CH_3 bonds can also be observed at 1250 cm^{-1} [17]. The presence of this band may be associated with a partial decomposition of the MMS molecule. In addition, the absorption band close to 620 cm^{-1} is related to the formation of tantalum suboxides due to oxidation and subsequent degradation of the tantalum wire [18]. The broadband close to 1600 cm^{-1} corresponds to water, due to the measurements being carried out in environmental conditions [12]. Figure 2b shows the absorbance spectra of the annealed films. In both cases, the absorption band related to CH_3 radicals is no longer present. In addition, a significative reduction in the band related to Si–O–H bonds for the $S3_{TA}$ film can be observed, while for the $S7_{TA}$ film, they do not appear after the thermal annealing.

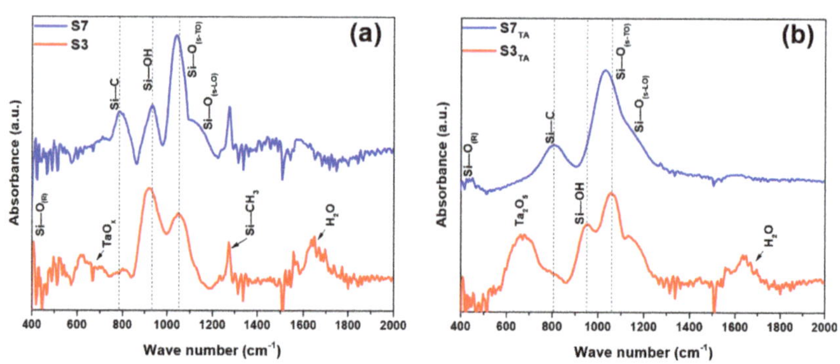

Figure 2. FTIR spectra of the SiO_xC_y:H (**a**) as-deposited films and (**b**) annealed films.

This behavior is related to a desorption process of hydrogen and CH_n radicals. Furthermore, the Si–O–Si stretching absorption band shows a shift to a higher wavenumber related to oxygen incorporation in the $S3_{TA}$ film. The absorption band related to silicon oxide shows a broadening and a shift from 1041 cm^{-1} to 1000 cm^{-1} for the $S7_{TA}$ film. These effects may be attributed to a greater amount of carbon-related bonds in the film. Many reports show that this broadband corresponds to the restructuring of silica materials in different link structures [19].

Figure 3a shows the photoluminescence (PL) spectra of the as-deposited films. The emission intensity for the film obtained at the highest MMS flow rate shows strong luminescence in the visible region, of about 4 times higher than for the film obtained at the lowest MMS flow rate. In addition, a broadening in the emission band and a red-shift in its maximum from 420 nm to 445 nm is observed when the MMS flow rate increases. Due to the material complexity, it is not possible to relate the luminescence origin with a single phenomenon. Among the different luminescent mechanisms are defects-related emissions, quantum confinement effects (QCE), and optical emissions in band tail states. In the case of defects-related emissions, these present emissions at specific wavelengths, so it is impossible to correlate the red-shift and the broadening observed in the emission spectra when the MMS flow rate increases with any particular defect [20]. However, both behaviors can be related to QCE and band tail states. On the one hand, the growth of Si/SiC nanocrystals could give rise to light emissions due to QCE where the increase in the nanocrystal sizes produces a red-shift in the emission wavelength [21]. The presence of nanocrystalline material could not be confirmed by XRD measurements or Raman spectroscopy; however, in a previous study using similar deposition parameters, we reported the obtention of polymorphous SiO_xC_y:H, that is, Si/SiC nanocrystalline phase embedded in an amorphous SiO_xC_y:H host matrix [22]. Considering the above, when the MMS flow increases, it is possible to induce the formation of larger nanocrystals with a broader size distribution, as suggested by the formation of larger clusters observed in the SEM micrographs. On the other hand, in an amorphous semiconductor, there is a continuum of possible states between the valence band and the conduction band, so the energy of the gap is not well defined. This induces the distribution of band tail states due to static disorder. The luminescence band is the result of the distribution of steady-state carriers within these states available for radiative recombination in the band tail [23]. In this case, when the carbon atoms replace the oxygen atoms, the band gap material tends to reduce its energy, producing the red-shift observed [1]. Moreover, the broadening in the emission band as a function of the carbon content may be related to the distribution of the density of band tail states, this is due to an increase in the structural disorder, where the substitution of the Si–Si bond with Si–C and C–C bonds takes place [24]. To have more elements to elucidate the origin of the luminescence in the films, the S3 and S7 films were thermally annealed with the conditions previously described. Figure 3b shows the PL spectra of the $S3_{TA}$ and $S7_{TA}$ films. In both cases, an increase in the emission intensity was observed after the thermal annealing. A remarkable change in the PL spectrum of the $S3_{TA}$ film was noticed. This spectrum displays a significative widening and shows two emission peaks centered in the violet and green regions. We can rule out the changes as a byproduct related to QCE. This is because the conditions in which the thermal annealing was carried out was not enough to induce changes in the fraction of nanocrystalline material, where at least temperatures of 1000 °C and an hour of annealing are required [25]. Nevertheless, the XPS and FTIR results suggest that the origin of these emission bands is related to the activation of luminescent defects in the SiO_xC_y:H matrix due to the desorption of hydrogen and other radicals. In this case, the emission band in the violet region can be associated with oxygen deficiency centers (ODCs), like E′ center, neutral oxygen vacancy (NOV), and the so-called C-related NOV state; while the emission band centered in the green region can be associated with the self-trapped exciton (STE), that is related to wide range luminescence spectra between 2 and 3 eV [26]. The STE is a transient defect, where the electron-hole pair generated after the light excitation can stay bonded by Coulomb forces, forming an exciton with a diameter close to the interatomic distances in Si–O bonds. Some models supports that a hole localized on oxygen and an electron localized on a neighboring silicon result in a small atomic displacement. Further displacement of the oxygen may induce the formation of permanent lattice defects as E′ centers and the non-bridging oxygen hole centers (NBOHC) [27]. Figure 3c shows a schematic diagram of the activation and formation of some of the radiative defects described above. The inset in Figure 3b shows the deconvoluted spectra of the $S3_{TA}$ film. We can observe a contribution from an emission band centered at 420 nm. This position is the same for the emission band in the as-deposited film, so we can assume that this emission band is related to the same emission mechanism. Considering the conditions for the thermal annealing, the results suggest that the origin of this emission band is related to QCE in the nanocrystalline fraction. On the other hand, for the $S7_{TA}$ film, the maximum of the PL spectra is centered in practically the same position as in the as-deposited film, and it displays a narrowing in the emission band. This may be related, on the one hand, to a reduction in the structural disorder of the film after the annealing induces a reduction in the width of the density distribution of the band tail states, causing a narrowing in the emission band; on the other hand, the emission band may be related to QCE by Si/SiC nanocrystals, since these should not show significant changes under this thermal annealing conditions. Figure 3d shows a schematic band diagram with the different recombination energy levels related to the luminescence in the SiO_xC_y:H films.

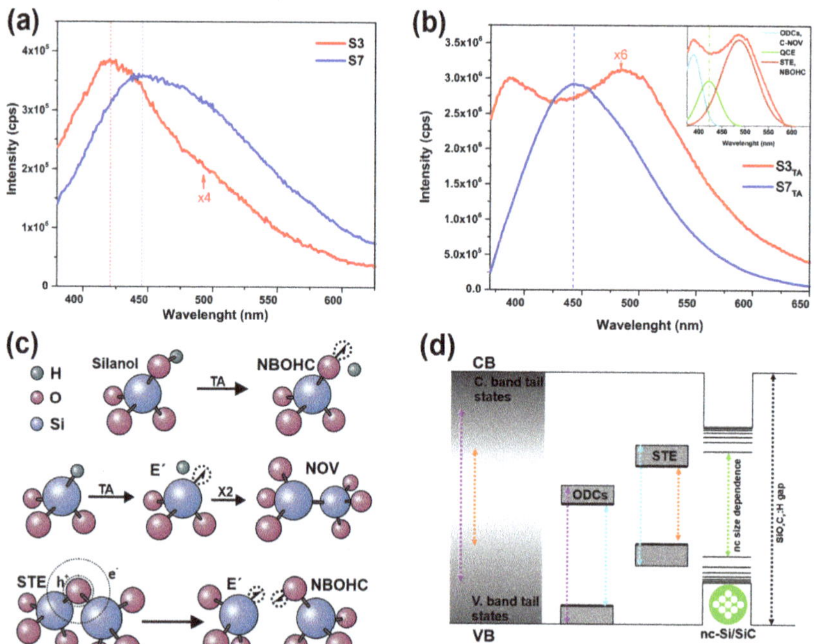

Figure 3. PL spectra of (**a**) as-deposited and (**b**) annealed SiO_xC_y:H films. (**c**) Schematic models of the activation and formation of radiative oxygen-deficient centers. (**d**) Band diagram for the recombination energy levels related to the luminescence in the SiO_xC_y:H films. In (**a**,**b**) dotted lines indicate the wavelength where is the maximum PL emission.

For the implementation of MOS-like electroluminescent structures only the annealed films were used due to these displayed higher PL intensities. Figure 4 shows the scanning electron microscopy (SEM) micrographs of the surface morphology and cross section of the $S3_{TA}$ and $S7_{TA}$ films. Both films show similar granular morphology with the formation of clusters on the surface of a slightly larger size for the $S7_{TA}$ film. A direct relation between the growth rate and the MMS flow was observed, where thicknesses of 40.8 nm and 100.2 nm were measured for samples $S3_{TA}$ and $S7_{TA}$, respectively.

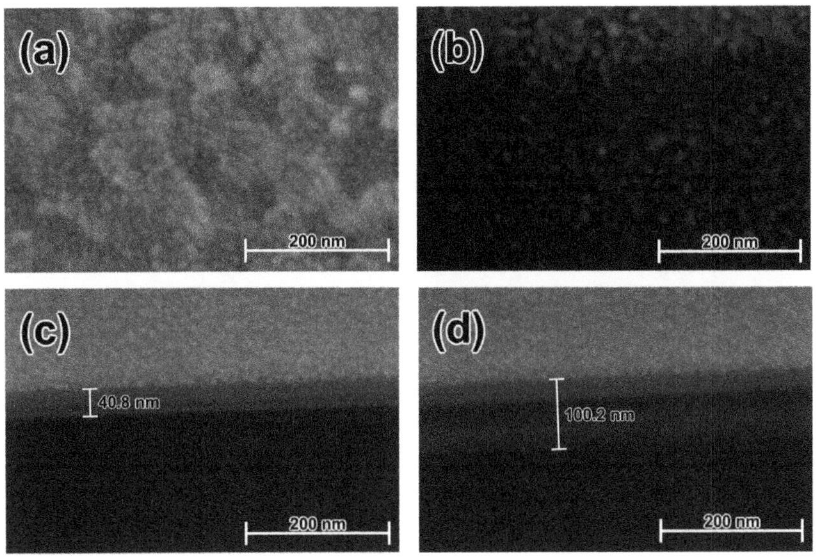

Figure 4. SEM micrographs of the annealed SiOxCy:H films. Surface morphology of (**a**) $S3_{TA}$ and (**b**) $S7_{TA}$. Cross section of (**c**) $S3_{TA}$ and (**d**) $S7_{TA}$.

2.2. *Electro-Optical Characterization of Structures*

To observe if the implemented structures display electroluminescence (EL), a voltage was applied both in forward bias (FB) and reverse bias (RB), limiting the current to a maximum of 1 mA to avoid damage to structures. Figure 5a shows the EL spectra of the $S3_{TA}$ and $S7_{TA}$ structures with FB voltages of 60 and 55 V, respectively. Only for the $S7_{TA}$ structure in FB, was it possible to see yellow light emissions in the entire area of the device with the naked eye in partial darkness, as we can observe in Figure 5b. The $S7_{TA}$ structure spectra show emissions in a broad band of the NIR-visible region, with a maximum emission close to 860 nm. Compared with the energy of the PL maximum, the EL maximum is notably less. Several authors have reported that in silicon-based materials where the emission is mainly related to QCE or radiative defects, when comparing the PL and EL spectra, they present emission bands with maximums in close positions [28,29]. In our case, the red-shift (~400 nm) could suggest that the EL observed is mainly related to band tail state emissions. Some models for luminescence in amorphous silicon-based materials mention that the red-shift observed is related to a thermalization of the electrons in the conduction band to inter-band tail states with lower energy [30]. Kruangam et al. suggested another reason for this decrease in the emission energy is that the holes are injected directly into the deep states of the gap [31]. The difference between the emission mechanisms of PL and EL is related to the additional process of the injection of carriers, since, while for PL the electron-hole pair is generated simultaneously at the same point in the active layer, the EL carriers are injected at different points of the structure, where their flow is also affected by their mobility in the material and the different barrier heights of the interface where they are injected. Because the mobility of the holes is lower, the carriers' recombination takes place near the c-Si/SiO_xC_y:H interface. This allows the electrons to thermalize, reducing emission energy. On the other hand, in both structures, intense and narrow emission peaks are observed in the blue region. This blue emission appears as bright spots on the surface of the structure, as can be seen in the images in Figure 5b,c. These narrow emission peaks are related to the characteristic radiation of indium ions due to the intense electric field applied to the ITO film [32].

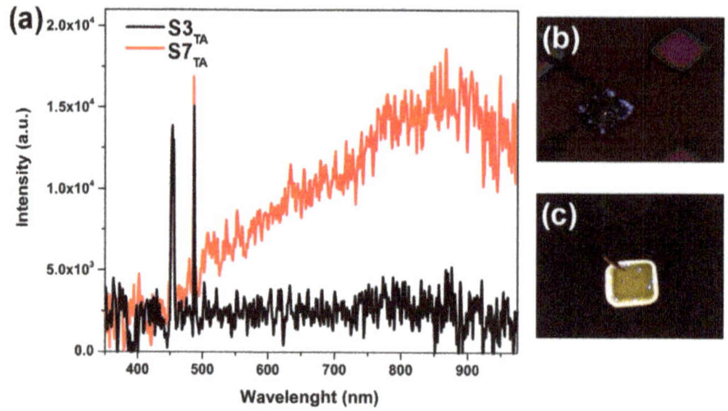

Figure 5. (**a**) Electroluminescence spectra of the $S3_{TA}$ and $S7_{TA}$ structures. (**b**,**c**) EL images of $S3_{TA}$ and $S7_{TA}$ structures, respectively.

To know the electrical properties of the structures, current density-voltage (J-V) characteristics were measured. Figure 6a shows the J-V characteristics in FB and RB of the $S3_{TA}$ and $S7_{TA}$. For both structures in FB, a higher current density is observed. However, the $S7_{TA}$ structure reaches higher current density values for the same FB voltage. This may be related first to a reduction in the gap energy, and then to an increase in the density of localized tail states; both phenomena are due to an increase in the carbon content in the active layer [33]. For low–medium FB voltages, similar current densities are observed in both structures; however, for the $S3_{TA}$ structure there is an abrupt drop in current density. This is related to the annihilation of preferential conductive paths. This is because these paths concentrate a greater current flow, also inducing an increase in temperature due to the Joule effect, causing structural changes for the breaking of Si–Si or Si–O bonds [34].

It is possible to elucidate the electrical charge transport mechanisms in the structure by analyzing the J-V characteristics under different conduction mechanism models. Within these models, those most related to this kind of structure are the Schottky emissions, direct tunneling, Fowler–Nordheim tunneling (F-N), Poole–Frenkel emissions (P-F), and trap-assisted tunneling (TAT) [35]. Only the structure that displayed EL ($S7_{TA}$) was analyzed under these models to identify the charge transport mechanisms. For the low–medium voltage region, the obtained J-V data fitted with the P-F and TAT models. Among the conduction mechanisms, the P-F emissions and TAT are considered as Bulk-limited conduction mechanisms, because the charge transport depends only on the electrical properties of the dielectric itself as the trap energy level, spacing, and density of these. P-F emissions occur in low–medium electric fields. This mechanism involves an electron in a trapping center in the dielectric film where this may be thermally excited out of the trap due to the reduction of the Coulomb potential energy of the electron by the applied electric field, enabling the conduction between localized states. The current density due to P-F emissions is given by Equation (1), where μ is the electronic drift mobility, ε_i and ε_0 are the dielectric constant and the permittivity in vacuum, respectively, N_c is the density of states in the conduction band, E is the electric field and the term ϕ_T is the trap energy level.

$$J_{P-F} = q\mu N_C E\, exp\left[\frac{-q\left(\phi_T - \sqrt{qE/\pi\varepsilon_i\varepsilon_0}\right)}{kT}\right] \quad (1)$$

As with P-F emissions, TAT tunneling involves an electron in a trapping center, where it may tunnel through the dielectric to reach the conduction band. The TAT mechanism also takes place in low–medium electric fields. The current density due to TAT emissions is given by Equation (2), where m_T^* is the tunneling effective mass in the dielectric and the other notations are the same as defined before.

$$J_{TAT} = exp\left[\frac{-8\pi\left(2qm_T^*\right)^{1/2}}{3hE}\phi_T^{3/2}\right] \quad (2)$$

Figure 6b shows the experimental J-V data fitted with the P-F and TAT models, according to Equations (1) and (2), respectively.

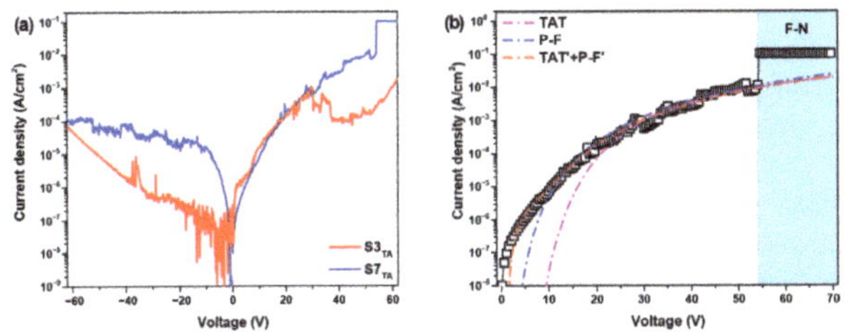

Figure 6. (**a**) Current density-voltage characteristics of the $S3_{TA}$ and $S7_{TA}$ structures in FB and RB. (**b**) J-V characteristic of the $S7_{TA}$ structure fitted to TAT and P-F conduction mechanism.

Both models show a fit with the J-V curve at the medium electric field region; however, the P-F model shows a better fit at the low electric field region. Considering that different conduction mechanisms can contribute to the conduction current through the dielectric simultaneously, a joint fit of both models was carried out, i.e., P-F′ + TAT′. We can observe a complete fit with the J-V curve. In this, two regions can be distinguished, the first for <10 V corresponds to P-F emissions, and the second for >10 V corresponds to the TAT. These results are related to a high density of localized tail states in the SiO_xC_y:H film. Due to the limited maximum current, it was impossible to identify the conduction mechanism for the high electric field region by comparing the J-V data with any model. However, in similar MOS-like structures, several authors report that the predominant conduction mechanism at high electric fields is F-N tunneling [36,37]. This mechanism is considered an electrode-limited conduction mechanism, where the barrier height at the electrode–dielectric interface is the most important parameter. F-N tunneling takes place when a high electric field induces the formation of a triangular potential barrier, narrow enough (<10 nm) for electrons to pass through into the conduction band of the dielectric because of the tunneling effect. Finally, based on the obtained results, Figure 7 shows the band diagram proposed for EL and the charge transport mechanism.

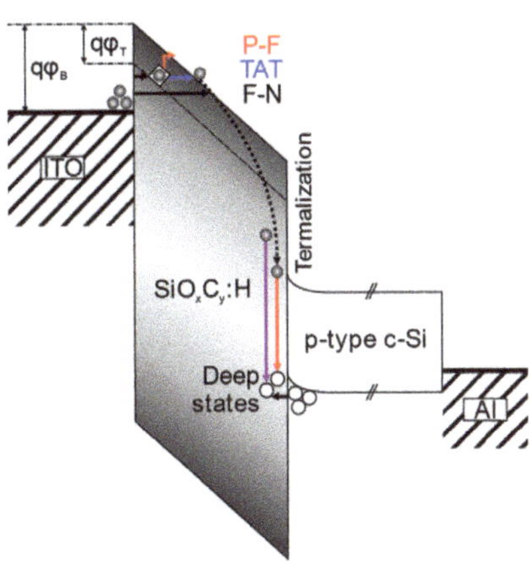

Figure 7. Schematic energy band diagram of the $S7_{TA}$ structure showing the EL and electrical conduction mechanism in FB condition.

3. Materials and Methods

3.1. Sample Preparation

Hydrogenated silicon oxycarbide (SiOxCy:H) thin films were deposited on p-type polished silicon substrates (100) with a resistivity of 1–4 Ω·cm by the HW-CVD technique. As catalyst material a 0.5 mm-diameter tantalum (Ta) wire at 1800 °C located 50 mm above the substrate was used. Hydrogen, oxygen, and MMS (purity 99.9%) were used as reactant gases. The hydrogen and oxygen flow rates were fixed at 20 sccm and 1 sccm, respectively, while the MMS flow rate was varied at 3 and 7 sccm (S3, S7). The films deposition maintained the substrate temperature at 200 °C and the chamber pressure at 0.1 Torr for 30 min. After deposition, the films were thermally annealed at 750 °C in nitrogen at atmospheric pressure for 30 min ($S3_{TA}$, $S7_{TA}$). After the characterization of the films, MOS-like structures were fabricated as electroluminescent devices using the annealed SiOxCy:H films as an active layer. A 100 nm thick indium thin oxide (ITO) film was deposited onto the $S3_{TA}$ and $S7_{TA}$ films by RF sputtering technique with a power of 60 W at room temperature, which was used as transparent contact. Then, 1 mm^2 square-shaped patterns were defined on the ITO contact. On the back side of the c-Si substrate, an approximately 1 μm thick aluminum layer was deposited by thermal evaporation. Finally, the structures were thermally annealed at 350 °C in a vacuum for 30 min to improve contact quality.

3.2. Characterization

The elemental composition of the films and the silicon chemical state were analyzed by X-ray photoelectron spectroscopy (XPS) in a system model K-Alpha (Thermo Scientific, Waltham, MA, USA) with an Al-Kα source. A Fourier-transform infrared spectrometer (FTIR) model Nicolet- iS50 (Thermo Scientific, Waltham, MA, USA), was used in a transmittance configuration in a range of 400–2000 cm^{-1} with a resolution of 4 cm^{-1}. The samples consisted of SiOxCy:H films deposited on c-Si substrates. Also, a clean c-Si substrate from the same wafer used for the films deposition was used to subtract the background signal from the absorbance spectra of the thin films. Photoluminescence (PL) spectra were obtained at room temperature in a Nanolog spectrometer from Horiba Jobin Yvon (Horiba Ltd., Kioto, Japan), using an excitation wavelength of 350 nm from a 450 W Xe-lamp. Surface morphology and cross-section micrographs were obtained by scanning electron microscopy (SEM) FEI Scios (Thermo Scientific, Waltham, MA, USA). To study the electro-optical properties of the SiO_xC_y:H based structures, the current density-voltage (J-V) characteristic was measured in a 4200-SCS parameter analyzer (Keithley, Solon, OH, USA), while the electroluminescence spectra were collected with an optical fiber coupled in a Horiba Duetta spectrometer (Horiba Ltd., Kioto, Japan).

4. Conclusions

The properties of SiO_xC_y:H thin films at different MMS flow rate were studied. A direct relation between the intensity and the width of the emission band with the carbon content in the films was observed. Also, a red-shift from 408 nm to 445 nm occurred when the MMS flow rate was increased. This shift and the broadening of the emission band suggest that luminescence is caused by optical transitions in band tail states with less contribution from QCEs in the nanocrystalline fraction. The annealed film $S7_{TA}$ showed a narrowing in the emission band; however, the emission maximum position remained the same. This suggests that the emission mechanisms are the same for the as-deposited film. On the other hand, the annealed film $S3_{TA}$ showed two emission bands centered in the violet and green regions, respectively. This was related to radiative defects in the SiO_xC_y:H matrix, due to the activation of different oxygen-deficient centers by the desorption of hydrogen and other radicals. Electroluminescent structures were implemented with the annealed films. Both structures displayed shiny blue dots on the surface related to the characteristic radiation of indium ions. Only the $S7_{TA}$ structure displayed an intense emissions in the entire area of yellow color when a 55 V FB was applied. The EL emission band showed a red-shift of about 400 nm with respect to the PL maximum. This is because the holes are injected into the deep band tail states and the electrons in the conduction band begin to lose energy as they move through the active layer due to the thermalization process. On the other hand, for this structure, the carriers conduction may be related first to the P-F emissions and the TAT due to a high density of band tail states, and second to F-N at high electric fields where the electrons can tunnel the triangular potential barrier to the conduction band.

Author Contributions: Conceptualization, J.R.R.-S. and Y.M.; methodology, J.R.R.-S., A.Á., G.R. and M.M.; validation A.Á., G.R. and M.M.; resources, J.R.R.-S., Y.M., A.M. and M.M.-M.; writing—original draft preparation, J.R.R.-S., Y.M., A.Á., G.R. and M.M.-M.; writing—review and editing, A.M., J.A.L., J.F. and G.M.M.; visualization, J.R.R.-S. and Y.M.; supervision, A.M., M.M.-M. and J.A.L.; project administration, J.R.R.-S., Y.M. and M.M.-M.; funding acquisition, J.R.R.-S., Y.M., J.A.L., M.M.-M., J.F. and G.M.M. All authors have read and agreed to the published version of the manuscript.

Funding: This research received no external funding.

Data Availability Statement: The original contributions presented in this study are included in the article. Further inquiries can be directed to the corresponding author.

Acknowledgments: J. R. Ramos-Serrano acknowledges CONAHCYT for the scholarship granted. The authors thank A. Mendez-Blas from IFUAP for PL measurements. We also appreciate Netzahualcoyotl Carlos Ramírez for SEM characterizations. We would like to express our sincere thanks to I. A. Oliva from the CINVESTAV-Mérida Unit for the XPS measurements. The authors thank Miguel A. Luna, Miguel Galván and Norma I. González from SEES-CINVESTAV for their technical assistance.

Conflicts of Interest: The authors declare no conflicts of interest.

References

1. Lin, Z.; Guo, Y.; Song, C.; Song, J.; Wang, X.; Zhang, Y.; Huang, R.; Huang, X. Influence of the Oxygen Content in Obtaining Tunable and Strong Photoluminescence from Low-Temperature Grown Silicon Oxycarbide Films. *J. Alloys Compd.* **2015**, *633*, 153–156. [CrossRef]
2. Baudzus, L.; Krummrich, P.M. Low-Loss Planar Optical Waveguides Based on Plasma Deposited Silicon Oxycarbide. *Opt. Mater. Express* **2019**, *9*, 2797. [CrossRef]
3. Matsumoto, Y. Hot Wire-CVD Deposited a-SiOx and Its Characterization. *Thin Solid Films* **2006**, *501*, 95–97. [CrossRef]
4. Nakayama, H.; Hata, T. Low-Temperature Growth of Si-Based Organic-Inorganic Hybrid Materials, Si-O-C and Si-N-C, by Organic Cat-CVD. *Thin Solid Films* **2006**, *501*, 190–194. [CrossRef]
5. Stannowski, B.; Rath, J.K.; Schropp, R.E.I. Growth Process and Properties of Silicon Nitride Deposited by Hot-Wire Chemical Vapor Deposition. *J. Appl. Phys.* **2003**, *93*, 2618–2625. [CrossRef]
6. Carlsson, J.-O.; Martin, P.M. Chemical Vapor Deposition. In *Handbook of Deposition Technologies for Films and Coatings*; Elsevier: Amsterdam, The Netherlands, 2010; pp. 314–363, ISBN 9780815520313.
7. Matsumoto, Y.; Godavarthi, S.; Ortega, M.; Sánchez, V.; Velumani, S.; Mallick, P.S. Size Modulation of Nanocrystalline Silicon Embedded in Amorphous Silicon Oxide by Cat-CVD. *Thin Solid Films* **2011**, *519*, 4498–4501. [CrossRef]
8. Finger, F.; Astakhov, O.; Bronger, T.; Carius, R.; Chen, T.; Dasgupta, A.; Gordijn, A.; Houben, L.; Huang, Y.; Klein, S.; et al. Microcrystalline Silicon Carbide Alloys Prepared with HWCVD as Highly Transparent and Conductive Window Layers for Thin Film Solar Cells. *Thin Solid Films* **2009**, *517*, 3507–3512. [CrossRef]
9. Benítez-Lara, A.; García-Salgado, G.; Vázquez-Valerdi, D.E.; Morales-Sánchez, A.; Espinosa-Torres, N.D.; Luna-López, J.A. Silicon Rich Oxide Powders by HWCVD: Its Optical and Morphological Properties. *Adv. Powder Technol.* **2015**, *26*, 163–168. [CrossRef]
10. Palacios-Huerta, L.; Cabañas-Tay, S.A.; Cardona-Castro, M.A.; Aceves-Mijares, M.; Domínguez-Horna, C.; Morales-Sánchez, A. Structural and Optical Properties of Silicon Rich Oxide Films in Graded-Stoichiometric Multilayers for Optoelectronic Devices. *Appl. Phys. Lett.* **2016**, *109*, 031906. [CrossRef]
11. Xu, H.; Wang, Q.; Xiao, H.; Li, X.; Su, X.; Tang, M.; Chen, L.; Li, S. In Situ Preparation of C-SiC x O y Coatings with Controllable Composition on Continuous Oxygen-Enriched SiC Fibres. *RSC Adv.* **2019**, *9*, 1319–1326. [CrossRef]
12. Basu, S. *Crystalline Silicon—Properties and Uses*; Basu, S., Ed.; IntechOpen: London, UK, 2011; ISBN 978-953-307-587-7.
13. Yang, C.S.; Yu, Y.H.; Lee, K.M.; Lee, H.J.; Choi, C.K. Investigation of Low Dielectric Carbon-Doped Silicon Oxide Films Prepared by PECVD Using Methyltrimethoxysilane Precursor. *Thin Solid Films* **2006**, *506–507*, 50–54. [CrossRef]
14. Carneiro, J.O.; Machado, F.; Rebouta, L.; Vasilevskiy, M.I.; Lanceros-Méndez, S.; Teixeira, V.; Costa, M.F.; Samantilleke, A.P. Compositional, Optical and Electrical Characteristics of SiOx Thin Films Deposited by Reactive Pulsed DC Magnetron Sputtering. *Coatings* **2019**, *9*, 468. [CrossRef]

15. Wang, S.; Ji, L.; Li, L.; Wu, Y.; Zhang, Y.; Lin, Z. Selective 6H-SiC White Light Emission by Picosecond Laser Direct Writing. *Sci Rep.* **2018**, *8*, 257. [CrossRef]
16. Costantini, J.-M.; Miro, S.; Pluchery, O. FTIR Study of Silicon Carbide Amorphization by Heavy Ion Irradiations. *J. Phys. D. Appl Phys.* **2017**, *50*, 095301. [CrossRef]
17. Kim, C.Y.; Lee, H.S.; Choi, C.K.; Yu, Y.H.; Navamathavan, R.; Lee, H.J. Method of Sealing Pores in Porous Low-k SiOC(-H) Films Fabricated Using Plasma-Assisted Atomic Layer Deposition. *J. Korean Phys. Soc.* **2013**, *62*, 1143–1149. [CrossRef]
18. Egorov, K.V.; Kuzmichev, D.S.; Sigarev, A.A.; Myakota, D.I.; Zarubin, S.S.; Chizov, P.S.; Perevalov, T.V.; Gritsenko, V.A.; Hwang C.S.; Markeev, A.M. Hydrogen Radical Enhanced Atomic Layer Deposition of TaOx: Saturation Studies and Methods for Oxygen Deficiency Control. *J. Mater. Chem. C* **2018**, *6*, 9667–9674. [CrossRef]
19. Xu, J.; Yang, C.S.; Choi, C.K. Annealing Effects on the Structural and Electrical Properties of SiOC(-H) Films with Low Dielectric Constant Prepared by Plasma-Enhanced Chemical Vapor Deposition. *J. Korean Phys. Soc.* **2004**, *45*, 175–179.
20. Salh, R. Defect Related Luminescence in Silicon Dioxide Network: A Review. In *Crystalline Silicon-Properties and Uses*; IntechOpen London, UK, 2011; Volume 135, p. 172. [CrossRef]
21. Dutt, A.; Salinas, R.A.; Martínez-Tolibia, S.E.; Ramos-Serrano, J.R.; Jain, M.; Hamui, L.; Ramos, C.D.; Mostafavi, E.; Kumar Mishra Y.; Matsumoto, Y.; et al. Silicon Compound Nanomaterials: Exploring Emission Mechanisms and Photobiological Applications *Adv. Photonics Res.* **2023**, *4*, 2300054. [CrossRef]
22. Ramos-Serrano, J.R.; Matsumoto, Y.; Méndez-Blas, A.; Dutt, A.; Morales, C.; Oliva, A.I. Luminescent Silicon Oxycarbide Thin Films Obtained with Monomethyl-Silane by Hot-Wire Chemical Vapor Deposition. *J. Alloys Compd.* **2019**, *780*, 341–346. [CrossRef]
23. Tessler, L.R.; Solomon, I. Photoluminescence of Tetrahedrally Coordinated A-Si1-XCx:H. *Phys. Rev. B* **1995**, *52*, 10962–10971 [CrossRef]
24. Güneş, İ.; Sel, K. Effects of Carbon Content and Plasma Power on Room Temperature Photoluminescence Characteristics of Hydrogenated Amorphous Silicon Carbide Thin Films Deposited by PECVD. *Thin Solid Films* **2017**, *636*, 85–92. [CrossRef]
25. Vivaldo, I.; Moreno, M.; Torres, A.; Ambrosio, R.; Rosales, P.; Carlos, N.; Calleja, W.; Monfil, K.; Benítez, A. A Comparative Study of Amorphous Silicon Carbide and Silicon Rich Oxide for Light Emission Applications. *J. Lumin.* **2017**, *190*, 215–220. [CrossRef]
26. Dutt, A.; Matsumoto, Y.; Santoyo-Salazar, J.; Santana-Rodríguez, G.; Godavarthi, S. Blue to Red Emission from As-Deposited Nc-Silicon/Silicon Dioxide by Hot Wire Chemical Vapor Deposition. *Thin Solid Films* **2015**, *595*, 221–225. [CrossRef]
27. Shluger, A.; Stefanovich, E. Models of the Self-Trapped Exciton and Nearest-Neighbor Defect Pair in SiO_2. *Phys. Rev. B* **1990**, *42*, 9664–9673. [CrossRef]
28. Walters, R.J.; Bourianoff, G.I.; Atwater, H.A. Field-Effect Electroluminescence in Silicon Nanocrystals. *Nat. Mater.* **2005**, *4*, 143–146 [CrossRef]
29. Cabañas-Tay, S.A.; Palacios-Huerta, L.; Aceves-Mijares, M.; Coyopol, A.; Morales-Morales, F.; Pérez-García, S.A.; Licea-Jiménez, L.; Domínguez-Horna, C.; Monfil-Leyva, K.; Morales-Sánchez, A. Study of Narrow and Intense UV Electroluminescence from ITO/SRO/Si-p and ITO/SRN/SRO/Si-p Based Light Emitting Capacitors. *J. Lumin.* **2017**, *183*, 334–340. [CrossRef]
30. Boulitrop, F.; Dunstan, D.J. Phonon Interactions in the Tail States of A-Si:H. *Phys. Rev. B* **1983**, *28*, 5923–5929. [CrossRef]
31. Kruangam, D.; Toyama, T.; Hattori, Y.; Deguchi, M.; Okamoto, H.; Hamakawa, Y. Improvement of Carrier Injection Efficiency in A-SiC p-i-n LED Using Highly-Conductive Wide-Gap p, n Type a-SiC Prepared by ECR CVD. *J. Non. Cryst. Solids* **1987**, *97–98*, 293–296. [CrossRef]
32. Cabanas-Tay, S.A.; Morales-Sanchez, A.; Palacios-Huerta, L.; Aceves-Mijares, M. UV Electroluminescence from ITO/SRO/p-Si and ITO/SRN/SRO/p-Si Structures. In Proceedings of the 2016 13th International Conference on Electrical Engineering, Computing Science and Automatic Control, CCE 2016, Mexico City, Mexico, 26–30 September 2016; IEEE: Piscataway, NJ, USA, 2016; pp. 1–5.
33. Sel, K.; Akaolu, B.; Atilgan, I.; Katirciolu, B. Effects of Tail States on the Conduction Mechanisms in Silicon Carbide Thin Films with High Carbon Content. *Solid. State. Electron.* **2011**, *57*, 1–8. [CrossRef]
34. Palacios-Huerta, L.; Cabanas-Tay, S.A.; Luna-López, J.A.; Aceves-Mijares, M.; Coyopol, A.; Morales-Sánchez, A. Effect of the Structure on Luminescent Characteristics of SRO-Based Light Emitting Capacitors. *Nanotechnology* **2015**, *26*, 395202. [CrossRef]
35. Chiu, F.C. A Review on Conduction Mechanisms in Dielectric Films. *Adv. Mater. Sci. Eng.* **2014**, *2014*, 578168. [CrossRef]
36. Ramírez, J.M.; Berencén, Y.; López-Conesa, L.; Rebled, J.M.; Peiró, F.; Garrido, B. Carrier Transport and Electroluminescence Efficiency of Erbium-Doped Silicon Nanocrystal Superlattices. *Appl. Phys. Lett.* **2013**, *103*, 081102. [CrossRef]
37. Lin, G.-R.; Lin, C.-J.; Kuo, H.-C. Improving Carrier Transport and Light Emission in a Silicon-Nanocrystal Based MOS Light-Emitting Diode on Silicon Nanopillar Array. *Appl. Phys. Lett.* **2007**, *91*, 093122. [CrossRef]

inorganics

Review

Preventing Dental Caries with Calcium-Based Materials: A Concise Review

Jieyi Chen [1,2], Yuqing Zhang [2], Iris Xiaoxue Yin [2], Ollie Yiru Yu [2], Alice Kit Ying Chan [2] and Chun Hung Chu [2,*]

1 Hospital of Stomatology, Sun Yat-sen University, Guangzhou 510055, China; chenjy679@mail.sysu.edu.cn
2 Faculty of Dentistry, The University of Hong Kong, Hong Kong SAR, China; monicayq@hku.hk (Y.Z.); irisxyin@hku.hk (I.X.Y.); ollieyu@hku.hk (O.Y.Y.); dralice@hku.hk (A.K.Y.C.)
* Correspondence: chchu@hku.hk

Abstract: This concise review provides an update on the use of calcium-based materials for the prevention of dental caries. Some calcium-based materials promote remineralization and neutralize bacterial acids, disrupting cariogenic biofilms and inhibiting bacterial growth. Medical Subject Headings of [Dental Caries] and [Calcium] were adopted to search publications. Information related to the aim of this review was extracted and summarized. Common calcium-based materials are calcium phosphate, hydroxyapatite, calcium carbonate, calcium fluoride and casein phosphopeptide–amorphous calcium phosphate (CPP-ACP). Calcium phosphate is commonly used in toothpaste. It provides calcium and phosphate ions, enhances the incorporation of fluoride into caries lesions and increases mineral density. Hydroxyapatite is a form of calcium phosphate that is chemically similar to the mineral found in teeth. It can be applied on teeth to prevent caries. Calcium carbonate can be found in toothpastes. It neutralizes bacterial acids and acts as a calcium reservoir during remineralization. Calcium fluoride is found in dental products and promotes remineralization as a source of fluoride, which can be incorporated into tooth enamel, forming fluorapatite and increasing resistance to caries. CPP-ACP is derived from milk proteins. It contains calcium and phosphate, which help to remineralize tooth enamel. CPP-ACP inhibits cariogenic bacteria. It also interacts with bacterial biofilms and disrupts their formation. These calcium-based materials can be used to boost the preventive effect of fluorides or, alternatively, as a therapy for caries prevention.

Keywords: caries; nanoparticles; calcium; phosphate; remineralization

1. Introduction

Dental caries is a chronic disease affecting over 2.3 billion permanent teeth and 532 million deciduous teeth worldwide, according to the Global Burden of Disease 2017 study report [1]. It is well-documented that dental caries is caused by the demineralization of hard tissues, including enamel, dentin and cementum, due to bacterial acids over time through a complex interaction between the host, fermentable carbohydrate and cariogenic bacteria [2]. Different agents have been developed to manage this common oral disease. The well-recognized foundation of dental caries prevention (primary and secondary) is fluoride-based agents, the effectiveness of which has been widely investigated and reported [3,4]. The major mechanism of fluoride is its reaction with enamel and dentine, along with calcium and phosphate, forming calcium–fluoride-like deposits during the process of remineralization [5]. Despite the well-known benefits of fluoride agents, dental researchers are exploring the use of different materials to prevent and arrest carious lesions. This may be due to the adverse effects of fluorides, like dental and skeletal fluorosis. The adverse effects also direct dental professionals to explore alternatives or boosters of fluorides for dental caries prevention [6].

Calcium-based materials have been introduced to prevent dental caries. The three types of hard tissues of the tooth, namely enamel, dentin and cementum, are mainly composed of hydroxyapatite, a mineral form of calcium. The net loss of calcium and phosphate ions from the tooth structure due to cariogenic bacteria will lead to dental caries [7]. Therefore, adding calcium and phosphate back to the tooth structure from supersaturated oral fluid and controlling bacteria activities plays an important role in dental caries prevention [8]. During the process of tooth remineralization, it is important that calcium and phosphate are abundant in the saliva. Unfortunately, however, calcium and phosphate are normally limited. Calcium-based materials refer to substances that contain calcium as a primary component or play a significant role in providing calcium ions. In the discipline of dental caries prevention, calcium-based materials are used to promote tooth remineralization and manage bacteria activity [9,10]. They can be used as adjunctive treatment with fluorides or alternative therapy for dental caries control. These materials show excellent biocompatibility, and some of them have been adopted in dental practice.

Understanding recent research advances helps clinical practitioners to incorporate the latest calcium-based materials and techniques into their practice and facilitate dental researchers in developing new formulations and improving this discipline. The aim of this narrative review is to provide an update on the use of calcium-based materials for dental caries prevention. The properties of different calcium-based materials, their ability to remineralize and manage bacterial activity and mechanisms, and their clinical applications and effectiveness are summarized herein.

2. Results

The results showed that calcium phosphate, hydroxyapatite, calcium carbonate, calcium fluoride and casein phosphopeptide–amorphous calcium phosphate (CPP-ACP) are the commonly reported calcium-based materials in the discipline of dental caries prevention. There are also other derivates of these calcium-based materials, such as sodium–calcium phosposilicate, sodium trimetaphosphate and calcium glucerophosphate as well as calcium nanoparticles. This manuscript mainly focuses on the commonly reported calcium-based materials (Table 1). Other forms of advanced materials are introduced in the discussion.

2.1. Calcium Phosphate

Calcium phosphates usually refer to a family of materials containing calcium ions combined with phosphate anions such as phosphate, monohydrogen phosphate and dihydrogen phosphate. The most frequently used ones are tricalcium phosphate and hydroxyapatite, which contains hydroxyl anions and is discussed in the following paragraph [11]. Tricalcium phosphate is commonly used in toothpaste, dentifrice, mouthwash and chewing gum to promote the remineralization of teeth [12]. It is a white solid which can be found in bone and tooth enamel. It is soluble in acid, but it has low solubility in water. It exists as three crystalline polymorphs, namely α, α' and β. Among them, β-tricalcium phosphate is stable at room temperature and becomes the bioavailable form used in oral care products. α-Tricalcium phosphate is less soluble than β-tricalcium phosphate. It is rare as it only exists at high temperatures above 1125 °C [13]. α'-Tricalcium phosphate is less common than α-tricalcium phosphate and has limited applications.

Researchers in the early 1990s found that the combined use of calcium phosphate and fluoride accelerates the setting and hardening of the compound [14]. However, it is difficult for these two substances to coexist in an oral care product as tricalcium phosphate can interact with fluoride to form calcium fluoride and therefore lower the bioavailability of both calcium and fluoride [15]. To solve the problem, a novel functionalized tricalcium phosphate was developed by coupling β-tricalcium phosphate with organic and/or inorganic moieties [16]. It was suggested that these molecules prevent premature fluoride–calcium interactions and enhance fluoride-based nucleation activity, subsequently leading to remineralization.

The potential of functionalized tricalcium phosphate for remineralization has been reported. In a laboratory study, functionalized tricalcium phosphate paste was applied on an artificial white spot lesion and showed a significantly greater increase in mean microhardness than casein phosphopeptide–amorphous calcium phosphate fluoride and the negative control [17]. Regarding enamel carious lesions, toothpaste containing functionalized tricalcium phosphate showed no significant difference in the percentage of remineralization potential and percentage of hardness recovery when compared to fluoride toothpaste [18]. A study reported that functionalized tricalcium phosphate promoted the incorporation of fluoride into root caries lesions and increased mineral density [19]. Few studies reported the effectiveness of functionalized tricalcium phosphate on cariogenic bacteria alone, given that it could be used in combination with other antibacterial agents. An in vitro study reported that the combination use of silver nitrate solution and sodium fluoride varnish containing functionalized tricalcium phosphate reduced cariogenic biofilm [20].

The clinical effectiveness of functionalized tricalcium phosphate in preventing dental caries has also been reported. In a clinical study, fluoride varnish was applied with functionalized tricalcium phosphate on white spot lesions at baseline and at 8 weeks among patients with fixed appliances and assessed at 16 weeks. The results indicated that 62% of the white spot lesions in the test group were reversed [21]. In another clinical study, silver nitrate solution and sodium fluoride varnish with functionalized tricalcium phosphate were adopted in order to arrest dentine caries among children [22]. The results indicated that the test group with functionalized tricalcium phosphate had a higher arrest rate than that without functionalized tricalcium phosphate. These results suggest that functionalized tricalcium phosphate has promising effects in preventing dental caries, but more clinical evidence is required to confirm its clinical use.

2.2. Hydroxyapatite

Hydroxyapatite is a naturally occurring mineral form of calcium apatite. Up to 70% by weight of human bone consists of a modified form of hydroxyapatite, and hydroxyapatite is the main mineral making up dental enamel and dentin. The solubility of hydroxyapatite in human dental enamel varies linearly with acidity, from 10 to 57 at pH 4.6 to 10 to 53 at pH 7.6 [23]. Hydroxyapatite is commonly added to toothpaste, dentifrice and mouthwash for daily oral care due to its ability to remineralize and manage oral bacteria [24]. A study reported that hydroxyapatite particles in toothpaste have the ability to bind to the damaged tooth surface and restore the surface integrity, and they are also able to penetrate into the deeper layers of the carious lesion [25]. Hydroxyapatite-based gel exhibited a significantly higher percentage of mineral gain in caries-like lesions after being subjected to the pH-cycling model when compared to artificial saliva [26]. Furthermore, hydroxyapatite in mouthwash could inhibit biofilm formation by preventing bacteria from binding to the tooth surface [27]. Instead, bacteria would bind to the hydroxyapatite particles and be cleared from the oral cavity. A review summarized that hydroxyapatite could be used to control biofilm by reducing bacterial attachment to enamel surfaces without antibacterial effects [28].

Some reviews and meta-analyses on the application of hydroxyapatite and its clinical effectiveness in preventing dental caries have been published. A systematic review on hydroxyapatite-based, fluoride-free oral care products was conducted and published in 2021. The results of the meta-analysis on three randomized control trials showed that hydroxyapatite toothpaste protected enamel from dental caries [29]. Another 18-month, double-blinded clinical trial reported that fluoride-free hydroxyapatite toothpaste was not statistically inferior to a fluoride toothpaste in preventing dental caries among adults [30]. The results of these studies suggest that hydroxyapatite is effective in preventing dental caries and has the potential to be an alternative to fluoride.

2.3. Calcium Carbonate

Calcium carbonate is the major constituent of limestone, marble, chalk, eggshells, bivalve shells and corals. It has the appearance of a white powder, a colorless crystal white powder or colorless crystals. Calcium carbonate is commonly added to dentifrice or toothpaste as an abrasive that removes dental plaque, and it is also used as a white coolant or thickener. It has low solubility in water but is soluble in dilute acid. Therefore, it has acid-buffering and calcium-releasing properties, which reverse the effects of the acids produced by bacteria [31].

It has been reported that calcium carbonate helps to neutralize plaque acid, increases the calcium level in dental plaque and acts as a calcium reservoir during remineralization [32]. A study reported that calcium–carbonate-based dentifrice was able to increase surface microhardness and was more effective in enamel remineralization when compared to the negative control [33]. Toothpaste containing arginine, calcium carbonate and sodium fluoride showed better recovery in microhardness and lesion depth after remineralization cycling treatment when compared to toothpaste containing sodium fluoride solution [34]. Adopting calcium carbonate alone to manage cariogenic bacteria was seldom reported on. An in vitro study reported that desensitizing paste containing 8% arginine and calcium carbonate reduced the formation of *Streptococcus mutans* biofilm on dentine [35].

As early as the 1950s, a study reported on the role of calcium carbonate in dental caries [36]. A clinical study reported that school children treated with a dentifrice containing regular fluoride content and calcium carbonate as a polishing agent had a tendency toward fewer decayed surfaces [37]. After that, several clinical trials reported that the adoption of calcium–carbonate-based fluoride toothpaste in daily oral health practice reduced dental caries increments [38,39]. An in vivo study reported that calcium–carbonate-based toothpastes could significantly reduce plaque acidity after a cariogenic challenge when compared to those without calcium carbonate [40]. These studies confirmed the clinical effectiveness of calcium carbonate in fluoridated dentifrice/toothpaste in preventing dental caries.

2.4. Calcium Fluoride

Calcium fluoride has a great variety of applications in dental caries prophylaxis. It is a white, solid, inorganic compound that is practically insoluble in water but slightly soluble in acid. The major cariostatic mechanism of fluoride is the formation of fluorapatite or calcium–fluoride-like precipitates, which are less soluble than hydroxyapatite when attacked by acid [41]. Both calcium and fluoride ions are prerequisites for remineralization, but premature calcium–fluoride interactions remain a challenge [42].

Calcium fluoride has also been suggested to act as an acidity-dependent reservoir of fluoride that can be incorporated into the tooth enamel, forming fluorapatite and increasing resistance to acid attacks [43]. However, its dissolution has been suggested to be controlled by the presence of phosphate and acidity. In a laboratory study, calcium fluoride particles were incubated with human enamel samples to provide a constant supply of soluble fluorides in a phosphate-free buffer solution. Despite this, the solubility of these particles and the release of fluoride were reduced when incubated in human saliva [44]. Though calcium fluoride has no bactericidal effect, it can be used in combination with other agents to enhance their antimicrobial properties. A laboratory study reported that composite materials modified with calcium fluoride highly reduced growth of two common cariogenic bacteria, *Lactobacillus acidophilus* and *Streptococcus mutans* [45]. Calcium fluoride can be used to produce gels, mouthwashes and pastes that strengthen tooth enamel and protect against acids. However, few clinical studies have been conducted that report the clinical effectiveness of calcium fluoride in preventing dental caries.

2.5. Casein Phosphopeptide–Amorphous Calcium Phosphate

Casein phosphopeptide–amorphous calcium phosphate (CPP-ACP) is commercially available and biocompatible and has attracted considerable research attention in recent years. Amorphous calcium phosphate (ACP) is a glassy solid. Casein phosphopeptides (CPPs) are derived from milk and can stabilize calcium and phosphate in ACP solution. This complex delivers calcium and phosphate ions to the tooth surface [46]. CPP-ACP can also interact with fluoride ions to form casein phosphopeptide–amorphous calcium fluoride phosphate (CPP-ACFP) [47].

CPP-ACP has been shown to have high remineralization potential for white spot lesions, dentine caries and non-cavitated root decay [48–50]. An in situ study revealed that fluoride toothpaste containing CPP-ACP increased the surface hardness of the demineralized enamel slab [51]. An in vitro study indicated that applying CPP-ACP on dentine surfaces led to lower demineralization and higher remineralization when immersed in demineralization solution [49]. It was proposed that ACP is localized in dental plaque and maintains a state of supersaturation with calcium and phosphate ions, thereby depressing demineralization and enhancing remineralization [52]. However, some reviewers reported that CPP-ACP/CPP-ACFP can inhibit demineralization but is less effective when compared to fluoride agents [53]. A systematic review with meta-analysis suggested that CPP-ACP/CPP-ACFP can be an adjunct to fluorides in caries prevention but cannot be an alternative [54].

CPP-ACP can also inhibit the activity of bacteria and the formation of biofilm. It has been reported that the addition of CPP-ACP to glass ionomer cement significantly reduced the growth of *Streptococcus mutans* in a dentine caries model [55]. Another laboratory study found that CPP-ACP reduces biofilm formation of *Streptococcus mutans* but does not kill the bacteria [56]. A clinical study reported that fluoride varnish containing CPP-ACP effectively reduced the salivary *Streptococcus mutans* count among children in mixed dentition [57]. Another clinical trial revealed that 5% sodium fluoride varnish with tricalcium phosphate significantly reduced saliva *mutans streptococci* among children with severe early-childhood caries [58].

CPP-ACP can be found in toothpaste, mouthwash, chewing gum and dental cream. A systematic review on clinical studies reported that fluorides combined with CPP-ACP achieved the same efficacy as fluorides monotherapy for early caries lesions on smooth surfaces, but the combination treatment had better efficacy than fluorides monotherapy for occlusal early carious lesions [59]. A randomized clinical trial reported that participants who chewed sugar-free gum containing CPP-ACP had 18% less tooth surface progression to dental caries when compared to those who chewed gum without CPP-ACP [60]. However, a systematic review also suggested that although CPP-ACP was more effective than a placebo in managing dental caries, its effectiveness was borderline when compared with fluoride [61]. Therefore, CPP-ACP can be an additive to fluoride but not an alternative to managing dental caries.

Table 1. Properties, application and clinical findings of calcium-based materials for caries prevention.

Calcium-Based Material	Properties	Applications	Clinical Findings [Reference]
Calcium phosphate	1. Remineralization	1. Toothpaste 2. Mouthwash 3. Chewing gum	1. Reverses white spot lesions [1] 2. Arrests caries with fluoride agents [21]
Hydroxyapatite	1. Remineralization 2. Bacterial inhibition	1. Toothpaste 2. Mouthwash	1. Reduces caries incidence [27,28]
Calcium carbonate	1. Remineralization 2. Buffer effect	1. Toothpaste	1. Reduces caries incidence [34] 2. Reduces caries increments [35,36]
Calcium fluoride	1. Remineralization 2. Bacterial inhibition	1. Mouthwash 2. Gel and paste	-
Casein phosphor–peptide–amorphous calcium phosphate	1. Remineralization 2. Bacterial inhibition	1. Toothpaste 2. Mouthwash 3. Chewing gum 4. Cream	1. Prevents early caries lesions [55] 2. Reduces caries incidence [56]

3. Methods

This concise review screened publications related to calcium-based materials in the discipline of dental caries prevention. Medical Subject Headings including [Dental Caries] and [Calcium] were adopted to search publications. Information regarding the properties of different calcium-based materials, their remineralization ability and impact on cariogenic bacteria and their clinical application was extracted and summarized.

4. Discussion

Calcium-based materials have shown great potential in the treatment of dental caries. Dental caries is a common dental problem caused by the demineralization of hard tissue by acids produced by cariogenic bacteria in the mouth. Therefore, the inhibition of demineralization and cariogenic bacteria is of great importance in preventing dental caries. During the process of remineralization, lost minerals such as calcium phosphate and fluoride are restored to the hard tissues, which helps to reverse the process of dental caries at the initial stage and inhibits further progression of cavitated lesions [12]. Calcium-based materials can saturate saliva and fluid with calcium ions (sometimes along with phosphate ions), neutralize the acidic environment, exchange ions between saliva and the tooth surface and form less soluble minerals [62]. In the past, the clinical use of calcium and phosphate ions for remineralization was not successful because of the low solubility of calcium-based materials as well as their premature interaction with fluoride ions [9]. In recent years, materials such as functionalized tricalcium phosphate have been developed to maintain the active calcium and fluoride during storage. Some studies suggested that calcium–phosphate materials had the potential to act as delivery systems in the mineralized tissue engineering field. These materials carry relevant substances to drive remineralization of dentin and/or stem cell differentiation within dental pulp [63]. New synthesis techniques may help to develop preferable calcium-based materials to prevent dental caries and reduce the risk of irreversible dental pulp damage in the future.

Some calcium-based materials have an influence on cariogenic bacteria. However, it has also been suggested that calcium-based materials do not have a bactericidal effect on their own. Instead, they inhibit the growth and adherence of cariogenic bacteria to control bacterial activity. CPP-ACP has been reported to interfere with the growth and adherence of *Streptococcus mutans* and *Streptococcus sobrinus* [64]. Free hydroxyapatite particles in toothpaste and mouthwash attract microorganisms, which attach to the particles and therefore clear microorganisms from the oral cavity [27]. A few studies reported the influence of calcium-based materials on the formation of biofilm even though these materials had no bactericide effect. These materials controlled the biofilm by inhibiting the attachment of bacteria to the tooth surface without killing bacteria. They can also be used in combination with other antibacterial agents and help to improve the effectiveness of controlling cariogenic bacteria and biofilm as well as the effectiveness of the prevention of dental caries. Though these calcium-based materials are widely used in daily oral health care, more clinical evidence is needed to confirm their clinical use and effectiveness in preventing dental caries among different people.

Nanotechnology has been adopted in the development of calcium-based materials such as nanoparticles of amorphous calcium phosphate, calcium phosphate nanocomposite and nano-hydroxyapatite. It has been reported that these calcium-based materials' nanoparticles release more ions than those at the micron scale because of the high surface-area-to-volume ratio [65]. As a result, the remineralization was enhanced. For example, nano-hydroxyapatite was reported to have greater remineralization effects on demineralized bovine dentine when compared to amine fluoride toothpastes [66]. A two-year clinical trial reported that nano-hydroxyapatite gel was effective in remineralizing proximal caries [67]. Nano-sized amorphous calcium phosphate was found to release more calcium and phosphate ions at a low pH, have lower mineral loss and decrease the carious lesion depth [68]. Thus, these nano-sized materials have the potential to be used in dental caries prevention, but more clinical evidence is needed in the future.

There are other derivates of these calcium-based materials, such as sodium–calcium phosposilicate, sodium trimetaphosphate, calcium glucerophosphate, glass–ionomer cement and bioactive glass. These materials have also been adopted in different forms of oral health care products to prevent dental caries. Toothpaste containing sodium–calcium phosposilicate has been reported to have the potential to remineralize enamel [69]. Fluoride varnishes containing sodium trimetaphosphate enhanced the remineralizing effect of artificial caries lesions by increasing the percentage of surface hardness recovery [70]. A literature review concluded that calcium glycerophosphate can reduce the plaque mass and have a greater anti-caries effect than sodium monofluorophosphate [71]. Glass ionomer cements, which are commonly used as a filling material and luting cement, had the ability to remineralize carious tooth surface and exert a bacteriostatic effect [72]. A randomized clinical trial indicated that glass ionomer sealants can effectively prevent dental caries [73]. A systematic review on bioactive glass, which can be adopted as a dental restorative material and mineralizing agent, indicated its effectiveness in remineralizing and forming apatite on the tooth surface of enamel and dentine. Besides, bioactive glass has also been reported to have an antibacterial effect on cariogenic bacteria [74].

Our concise review summarizes the commonly used calcium-based materials and provides an overview of their properties as well as their role and clinical effectiveness in dental caries prevention. Nevertheless, this concise review provides an overview and not in-depth details for researchers.

5. Conclusions

Calcium-based materials such as calcium phosphate, hydroxyapatite, calcium carbonate, calcium fluoride and casein phosphopeptide–amorphous calcium phosphate can be used to boost the preventive effect of fluorides or, alternatively, as a therapy for caries prevention. Moreover, they have the potential to be used as drug delivery systems in the mineralized tissue engineering field. More clinical evidence is needed to warrant their clinical use.

Author Contributions: Conceptualization, J.C. and C.H.C.; methodology, J.C. and Y.Z.; resources, C.H.C.; data curation, J.C. and C.H.C.; writing—original draft preparation, J.C., Y.Z. and I.X.Y.; writing—review and editing, O.Y.Y., A.K.Y.C. and C.H.C.; supervision, C.H.C.; funding acquisition, C.H.C. All authors have read and agreed to the published version of the manuscript.

Funding: This review was funded by UGC General Research Fund 17100222.

Data Availability Statement: This concise review has no database.

Conflicts of Interest: The authors declare no conflicts of interest.

References

1. GBD 2017 Disease and Injury Incidence and Prevalence Collaborators. Global, regional, and national incidence, prevalence, and years lived with disability for 354 diseases and injuries for 195 countries and territories, 1990–2017: A systematic analysis for the Global Burden of Disease Study 2017. *Lancet* **2018**, *392*, 1789–1858. [CrossRef] [PubMed]
2. Selwitz, R.H.; Ismail, A.I.; Pitts, N.B. Dental caries. *Lancet* **2007**, *369*, 51–59. [CrossRef] [PubMed]
3. Featherstone, J.D. Prevention and reversal of dental caries: Role of low level fluoride. *Community Dent. Oral Epidemiol.* **1999**, *27*, 31–40. [CrossRef] [PubMed]
4. Horst, J.A.; Heima, M. Prevention of Dental Caries by Silver Diamine Fluoride. *Compend. Contin. Educ. Dent.* **2019**, *40*, 158–163. [PubMed]
5. Pollick, H. The Role of Fluoride in the Prevention of Tooth Decay. *Pediatr. Clin. N. Am.* **2018**, *65*, 923–940. [CrossRef]
6. Veneri, F.; Iamandii, I.; Vinceti, M.; Birnbaum, L.S.; Generali, L.; Consolo, U.; Filippini, T. Fluoride Exposure and Skeletal Fluorosis: A Systematic Review and Dose-response Meta-analysis. *Curr. Environ. Health Rep.* **2023**, *10*, 417–441. [CrossRef]
7. Kutsch, V.K. Dental caries: An updated medical model of risk assessment. *J. Prosthet. Dent.* **2014**, *111*, 280–285. [CrossRef]
8. Abou Neel, E.A.; Aljabo, A.; Strange, A.; Ibrahim, S.; Coathup, M.; Young, A.M.; Bozec, L.; Mudera, V. Demineralization-remineralization dynamics in teeth and bone. *Int. J. Nanomed.* **2016**, *11*, 4743–4763. [CrossRef]
9. Reynolds, E.C. Calcium phosphate-based remineralization systems: Scientific evidence? *Aust. Dent. J.* **2008**, *53*, 268–273. [CrossRef]

10. Schlafer, S.; Birkedal, H.; Olsen, J.; Skovgaard, J.; Sutherland, D.S.; Wejse, P.L.; Nyvad, B.; Meyer, R.L. Calcium-phosphate-osteopontin particles for caries control. *Biofouling* **2016**, *32*, 349–357. [CrossRef]
11. Al-Sanabani, J.S.; Madfa, A.A.; Al-Sanabani, F.A. Application of calcium phosphate materials in dentistry. *Int. J. Biomater.* **2013** *2013*, 876132. [CrossRef] [PubMed]
12. Limeback, H.; Enax, J.; Meyer, F. Improving Oral Health with Fluoride-Free Calcium-Phosphate-Based Biomimetic Toothpastes An Update of the Clinical Evidence. *Biomimetics* **2023**, *8*, 331. [CrossRef] [PubMed]
13. Dorozhkin, S.V.; Epple, M. Biological and medical significance of calcium phosphates. *Angew. Chem. Int. Ed. Engl.* **2002**, *41*, 3130–3146. [CrossRef] [PubMed]
14. Mirtchi, A.A.; Lemaître, J.; Munting, E. Calcium phosphate cements: Effect of fluorides on the setting and hardening of β-tricalcium phosphate-dicalcium phosphate-calcite cements. *Biomaterials* **1991**, *12*, 505–510. [CrossRef] [PubMed]
15. Walsh, L.J. Contemporary technologies for remineralization therapies: A review. *Int. Dent. SA* **2009**, *11*, 6–16.
16. Karlinsey, R.L.; Pfarrer, A.M. Fluoride plus functionalized β-TCP: A promising combination for robust remineralization. *Adv. Dent. Res.* **2012**, *24*, 48–52. [CrossRef]
17. Bhadoria, N.; Gunwal, M.K.; Kukreja, R.; Maran, S.; Devendrappa, S.N.; Singla, S. An In Vitro Evaluation of Remineralization Potential of Functionalized Tricalcium Phosphate Paste and CPP-ACPF on Artificial White Spot Lesion in Primary and Permanent Enamel. *Int. J. Clin. Pediatr. Dent.* **2020**, *13*, 579–584.
18. Juntavee, A.; Juntavee, N.; Hirunmoon, P. Remineralization Potential of Nanohydroxyapatite Toothpaste Compared with Tricalcium Phosphate and Fluoride Toothpaste on Artificial Carious Lesions. *Int. J. Dent.* **2021**, *2021*, 5588832. [CrossRef]
19. Cai, J.; Burrow, M.F.; Manton, D.J.; Hardiman, R.; Palamara, J.E.A. Remineralising effects of fluoride varnishes containing calcium phosphate on artificial root caries lesions with adjunctive application of proanthocyanidin. *Dent. Mater.* **2021**, *37*, 143–157. [CrossRef]
20. Yu, O.Y.; Zhao, I.S.; Mei, M.L.; Lo, E.C.; Chu, C.H. Effect of Silver Nitrate and Sodium Fluoride with Tri-Calcium Phosphate on Streptococcus mutans and Demineralised Dentine. *Int. J. Mol. Sci.* **2018**, *19*, 1288. [CrossRef]
21. Salamara, O.; Papadimitriou, A.; Mortensen, D.; Twetman, S.; Koletsi, D.; Gizani, S. Effect of fluoride varnish with functionalized tri-calcium phosphate on post-orthodontic white spot lesions: An investigator-blinded controlled trial. *Quintessence Int.* **2020**, *51*, 854–862. [PubMed]
22. Chen, K.J.; Gao, S.S.; Duangthip, D.; Lo, E.C.M.; Chu, C.H. Randomized Clinical Trial on Sodium Fluoride with Tricalcium Phosphate. *J. Dent. Res.* **2021**, *100*, 66–73. [CrossRef] [PubMed]
23. Larsen, M.J.; Jensen, S.J. The hydroxyapatite solubility product of human dental enamel as a function of pH in the range 4.6-7.6 at 20 degrees C. *Arch. Oral Biol.* **1989**, *34*, 957–961. [CrossRef] [PubMed]
24. Pawinska, M.; Paszynska, E.; Limeback, H.; Amaechi, B.T.; Fabritius, H.-O.; Ganss, B.; O'hagan-Wong, K.; Wiesche, E.S.Z.; Meyer, F.; Enax, J. Hydroxyapatite as an active ingredient in oral care: An international symposium report. *Bioinspired Biomim. Nanobiomater.* **2024**, *13*, 1–14. [CrossRef]
25. O'Hagan-Wong, K.; Enax, J.; Meyer, F.; Ganss, B. The use of hydroxyapatite toothpaste to prevent dental caries. *Odontology* **2022**, *110*, 223–230. [CrossRef]
26. Amaechi, B.T.; AbdulAzees, P.A.; Okoye, L.O.; Meyer, F.; Enax, J. Comparison of hydroxyapatite and fluoride oral care gels for remineralization of initial caries: A pH-cycling study. *BDJ Open* **2020**, *6*, 9. [CrossRef]
27. Kensche, A.; Holder, C.; Basche, S.; Tahan, N.; Hannig, C.; Hannig, M. Efficacy of a mouthrinse based on hydroxyapatite to reduce initial bacterial colonisation in situ. *Arch. Oral Biol.* **2017**, *80*, 18–26. [CrossRef]
28. Meyer, F.; Enax, J. Hydroxyapatite in Oral Biofilm Management. *Eur. J. Dent.* **2019**, *13*, 287–290. [CrossRef]
29. Limeback, H.; Enax, J.; Meyer, F. Biomimetic hydroxyapatite and caries prevention: A systematic review and meta-analysis. *Can. J. Dent. Hyg.* **2021**, *55*, 148–159.
30. Paszynska, E.; Pawinska, M.; Enax, J.; Meyer, F.; Schulze Zur Wiesche, E.; May, T.W.; Amaechi, B.T.; Limeback, H.; Hernik, A.; Otulakowska-Skrzynska, J.; et al. Caries-preventing effect of a hydroxyapatite-toothpaste in adults: A 18-month double-blinded randomized clinical trial. *Front. Public Health* **2023**, *11*, 1199728. [CrossRef]
31. Meyer, F.; Schulze Zur Wiesche, E.; Amaechi, B.T.; Limeback, H.; Enax, J. Caries Etiology and Preventive Measures. *Eur. J. Dent.* **2024**, *18*, 766–776. [CrossRef] [PubMed]
32. Lynch, R.J.; ten Cate, J.M. The anti-caries efficacy of calcium carbonate-based fluoride toothpastes. *Int. Dent. J.* **2005**, *55* (Suppl. 1), 175–178. [CrossRef] [PubMed]
33. Cury, J.A.; Simões, G.S.; Del Bel Cury, A.A.; Gonçalves, N.C.; Tabchoury, C.P. Effect of a calcium carbonate-based dentifrice on in situ enamel remineralization. *Caries Res.* **2005**, *39*, 255–257. [CrossRef] [PubMed]
34. Huang, Y.; Duan, Y.; Qian, Y.; Huang, R.; Yang, Z.; Li, Y.; Zhou, Z. Remineralization efficacy of a toothpaste containing 8% arginine and calcium carbonate on enamel surface. *Am. J. Dent.* **2013**, *26*, 291–297. [PubMed]
35. Fu, D.; Pei, D.; Huang, C.; Liu, Y.; Du, X.; Sun, H. Effect of desensitising paste containing 8% arginine and calcium carbonate on biofilm formation of Streptococcus mutans in vitro. *J. Dent.* **2013**, *41*, 619–627. [CrossRef] [PubMed]
36. Gore, J.T. The role of calcium carbonate in dental caries. *J. Am. Dent. Assoc.* **1953**, *47*, 180–189. [CrossRef]
37. Forsman, B. Studies on the effect of dentifrices with low fluoride content. *Community Dent. Oral Epidemiol.* **1974**, *2*, 166–175. [CrossRef]

38. Naylor, M.N.; Glass, R.L. A 3-year clinical trial of calcium carbonate dentifrice containing calcium glycerophosphate and sodium monofluorophosphate. *Caries Res.* **1979**, *13*, 39–46. [CrossRef]
39. Mainwaring, P.J.; Naylor, M.N. A four-year clinical study to determine the caries-inhibiting effect of calcium glycerophosphate and sodium fluoride in calcium carbonate base dentifrices containing sodium monofluorophosphate. *Caries Res.* **1983**, *17*, 267–276. [CrossRef]
40. Tahmassebi, J.; Duggal, M.S.; Curzon, M.E. Effect of a calcium carbonate-based toothpaste with 0.3% triclosan on pH changes in dental plaque in vivo. *Caries Res.* **1994**, *28*, 272–276. [CrossRef]
41. Rošin-Grget, K.; Peroš, K.; Sutej, I.; Bašić, K. The cariostatic mechanisms of fluoride. *Acta Med. Acad.* **2013**, *42*, 179–188. [CrossRef] [PubMed]
42. Kantrong, N.; Khongkhaphet, K.; Sitornsud, N.; Lo-Apirukkul, P.; Phanprom, W.; Rojviriya, C.; Amonpattaratkit, P.; Ariyakriangkai, W. Synchrotron radiation analysis of root dentin: The roles of fluoride and calcium ions in hydroxyapatite remineralization. *J. Synchrotron Radiat.* **2022**, *29*, 496–504. [CrossRef] [PubMed]
43. Ogaard, B. CaF_2 formation: Cariostatic properties and factors of enhancing the effect. *Caries Res.* **2001**, *35* (Suppl. 1), 40–44. [CrossRef] [PubMed]
44. Koeser, J.; Carvalho, T.S.; Pieles, U.; Lussi, A. Preparation and optimization of calcium fluoride particles for dental applications. *J. Mater. Sci. Mater. Med.* **2014**, *25*, 1671–1677. [CrossRef] [PubMed]
45. Łukomska-Szymańska, M.; Zarzycka, B.; Grzegorczyk, J.; Sokołowski, K.; Półtorak, K.; Sokołowski, J.; Łapińska, B. Antibacterial Properties of Calcium Fluoride-Based Composite Materials: In Vitro Study. *BioMed Res. Int.* **2016**, *2016*, 1048320.
46. Reynolds, E.C. Remineralization of enamel subsurface lesions by casein phosphopeptide-stabilized calcium phosphate solutions. *J. Dent. Res.* **1997**, *76*, 1587–1595. [CrossRef]
47. Cross, K.J.; Huq, N.L.; Palamara, J.E.; Perich, J.W.; Reynolds, E.C. Physicochemical characterization of casein phosphopeptide-amorphous calcium phosphate nanocomplexes. *J. Biol. Chem.* **2005**, *280*, 15362–15369. [CrossRef]
48. Indrapriyadharshini, K.; Madan Kumar, P.D.; Sharma, K.; Iyer, K. Remineralizing potential of CPP-ACP in white spot lesions—A systematic review. *Indian. J. Dent. Res.* **2018**, *29*, 487–496. [CrossRef]
49. Rahiotis, C.; Vougiouklakis, G. Effect of a CPP-ACP agent on the demineralization and remineralization of dentine in vitro. *J. Dent.* **2007**, *35*, 695–698. [CrossRef]
50. Mohammadi, N.; Rikhtegaran, S.; Kimyai, S.; Rahbar, M.; Pirzadeh, T.; Asdagh, S.; Sezevar, A. The Effect of Photodynamic Therapy and Casein Phosphopeptide-Amorphous Calcium Phosphate (CPP-ACP) on the Remineralization Rate of Non-Cavitated Root: An In-vitro Study. *Maedica* **2019**, *14*, 357–362. [CrossRef]
51. de Oliveira, P.R.A.; Barreto, L.; Tostes, M.A. Effectiveness of CPP-ACP and fluoride products in tooth remineralization. *Int. J. Dent. Hyg.* **2022**, *20*, 635–642. [CrossRef] [PubMed]
52. Jefferies, S.R. Advances in remineralization for early carious lesions: A comprehensive review. *Compend. Contin. Educ. Dent.* **2014**, *35*, 237–243; quiz 244. [PubMed]
53. Reise, M.; Kranz, S.; Heyder, M.; Jandt, K.D.; Sigusch, B.W. Effectiveness of Casein Phosphopeptide-Amorphous Calcium Phosphate (CPP-ACP) Compared to Fluoride Products in an In-Vitro Demineralization Model. *Materials* **2021**, *14*, 5974. [CrossRef] [PubMed]
54. Bijle, M.N.A.; Yiu, C.K.Y.; Ekambaram, M. Calcium-Based Caries Preventive Agents: A Meta-evaluation of Systematic Reviews and Meta-analysis. *J. Evid. Based Dent. Pract.* **2018**, *18*, 203–217.e4. [CrossRef] [PubMed]
55. Pinheiro, S.L.; Azenha, G.R.; De Milito, F.; Democh, Y.M. Antimicrobial Capacity of Casein Phosphopeptide/Amorphous Calcium Phosphate and Enzymes in Glass Ionomer Cement in Dentin Carious Lesions. *Acta Stomatol. Croat.* **2015**, *49*, 104–111. [CrossRef]
56. Sionov, R.V.; Tsavdaridou, D.; Aqawi, M.; Zaks, B.; Steinberg, D.; Shalish, M. Tooth mousse containing casein phosphopeptide-amorphous calcium phosphate prevents biofilm formation of Streptococcus mutans. *BMC Oral Health* **2021**, *21*, 136. [CrossRef]
57. Patel, P.M.; Hugar, S.M.; Halikerimath, S.; Badakar, C.M.; Gokhale, N.S.; Thakkar, P.J.; Kohli, D.; Shah, S. Comparison of the Effect of Fluoride Varnish, Chlorhexidine Varnish and Casein Phosphopeptide-Amorphous Calcium Phosphate (CPP-ACP) Varnish on Salivary Streptococcus mutans Level: A Six Month Clinical Study. *J. Clin. Diagn. Res.* **2017**, *11*, ZC53–ZC59. [CrossRef]
58. Erkmen Almaz, M.; Akbay Oba, A. Antibacterial activity of fluoride varnishes containing different agents in children with severe early childhood caries: A randomised controlled trial. *Clin. Oral Investig.* **2020**, *24*, 2129–2136. [CrossRef]
59. Tao, S.; Zhu, Y.; Yuan, H.; Tao, S.; Cheng, Y.; Li, J.; He, L. Efficacy of fluorides and CPP-ACP vs fluorides monotherapy on early caries lesions: A systematic review and meta-analysis. *PLoS ONE* **2018**, *13*, e0196660. [CrossRef]
60. Morgan, M.V.; Adams, G.G.; Bailey, D.L.; Tsao, C.E.; Fischman, S.L.; Reynolds, E.C. The anticariogenic effect of sugar-free gum containing CPP-ACP nanocomplexes on approximal caries determined using digital bitewing radiography. *Caries Res.* **2008**, *42*, 171–184. [CrossRef]
61. Wang, Y.; Li, J.; Sun, W.; Li, H.; Cannon, R.D.; Mei, L. Effect of non-fluoride agents on the prevention of dental caries in primary dentition: A systematic review. *PLoS ONE* **2017**, *12*, e0182221. [CrossRef] [PubMed]
62. Cochrane, N.J.; Cai, F.; Huq, N.L.; Burrow, M.F.; Reynolds, E.C. New approaches to enhanced remineralization of tooth enamel. *J. Dent. Res.* **2010**, *89*, 1187–1197. [CrossRef] [PubMed]
63. Cuylear, D.L.; Elghazali, N.A.; Kapila, S.D.; Desai, T.A. Calcium Phosphate Delivery Systems for Regeneration and Biomineralization of Mineralized Tissues of the Craniofacial Complex. *Mol. Pharm.* **2023**, *20*, 810–828. [CrossRef] [PubMed]

64. Rose, R.K. Binding characteristics of Streptococcus mutans for calcium and casein phosphopeptide. *Caries Res.* **2000**, *34*, 427–431. [CrossRef] [PubMed]
65. Rodrigues, M.C.; Natale, L.C.; Arana-Chaves, V.E.; Braga, R.R. Calcium and phosphate release from resin-based materials containing different calcium orthophosphate nanoparticles. *J. Biomed. Mater. Res. B Appl. Biomater.* **2015**, *103*, 1670–1678. [CrossRef]
66. Tschoppe, P.; Zandim, D.L.; Martus, P.; Kielbassa, A.M. Enamel and dentine remineralization by nano-hydroxyapatite toothpastes. *J. Dent.* **2011**, *39*, 430–437. [CrossRef]
67. Grocholewicz, K.; Matkowska-Cichocka, G.; Makowiecki, P.; Droździk, A.; Ey-Chmielewska, H.; Dziewulska, A.; Tomasik, M.; Trybek, G.; Janiszewska-Olszowska, J. Effect of nano-hydroxyapatite and ozone on approximal initial caries: A randomized clinical trial. *Sci. Rep.* **2020**, *10*, 11192. [CrossRef]
68. Melo, M.A.S.; Weir, M.D.; Passos, V.F.; Powers, M.; Xu, H.H.K. Ph-activated nano-amorphous calcium phosphate-based cement to reduce dental enamel demineralization. *Artif. Cells Nanomed. Biotechnol.* **2017**, *45*, 1778–1785. [CrossRef]
69. Gjorgievska, E.S.; Nicholson, J.W. A preliminary study of enamel remineralization by dentifrices based on RECALDENTTM (CPP-ACP) and Novamin®(calcium-sodium-phosphosilicate). *Acta Odontol. Latinoam.* **2010**, *23*, 234–239.
70. Manarelli, M.; Delbem, A.C.B.; Lima, T.; Castilho, F.; Pessan, J.P. In vitro remineralizing effect of fluoride varnishes containing sodium trimetaphosphate. *Caries Res.* **2014**, *48*, 299–305. [CrossRef]
71. Lynch, R. Calcium glycerophosphate and caries: A review of the literature. *Int. Dent. J.* **2004**, *54*, 310–314. [CrossRef] [PubMed]
72. Ge, K.X.; Quock, R.; Chu, C.H.; Yu, O.Y. The preventive effect of glass ionomer cement restorations on secondary caries formation: A systematic review and meta-analysis. *Dent. Mater.* **2023**, *39*, e1–e17. [CrossRef] [PubMed]
73. Barja-Fidalgo, F.; Maroun, S.; de Oliveira, B.H. Effectiveness of a glass ionomer cement used as a pit and fissure sealant in recently erupted permanent first molars. *J. Dent. Child.* **2009**, *76*, 34–40.
74. Dai, L.L.; Mei, M.L.; Chu, C.H.; Lo, E.C.M. Mechanisms of Bioactive Glass on Caries Management: A Review. *Materials* **2019**, *12*, 4183. [CrossRef]

inorganics MDPI

Review

Progress, Applications, and Challenges of Amorphous Alloys: A Critical Review

Zheyuan Feng [1], Hansheng Geng [1], Yuze Zhuang [2] and Pengwei Li [3,*]

1 State Key Laboratory of Disaster Prevention & Mitigation of Explosion & Impact, Army Engineering University of PLA, Nanjing 210007, China
2 Suzhou High Speed Management Co., Ltd., Suzhou 215021, China
3 Key Laboratory of Superlight Materials and Surface Technology of Ministry of Education, College of Materials Science and Chemical Engineering, Harbin Engineering University, Harbin 150001, China
* Correspondence: pengweili@bio.aau.dk

Abstract: Amorphous alloys, also known as metallic glasses, are a type of novel amorphous material discovered by chance. This discovery has greatly enriched the field of metal physics, spurred the rapid development of amorphous physics and materials science, and propelled amorphous physics to the forefront of condensed matter physics. As an important and challenging branch of this discipline, amorphous physics now plays a pivotal role in understanding the complexities of non-crystalline materials. Amorphous materials, characterized by their unique properties, are not only widely used in daily life and high-tech fields but also serve as model systems for studying significant scientific issues within materials science and condensed matter physics. This paper provides a comprehensive review of amorphous alloys, discussing major scientific issues and challenges in amorphous science, the formation mechanisms of these materials, their structural characteristics, and their physical and mechanical properties. Additionally, it explores the various applications of amorphous materials and forecasts future research trends, significant issues, development prospects, and directions within this vibrant field.

Keywords: amorphous alloys; microcosmic; mechanical properties; physical and chemical properties

1. Introduction

Amorphous alloys, also known as metallic glasses or liquid metals, are novel materials synthesized using modern rapid solidification techniques [1]. These alloys combine the excellent mechanical, physical, and chemical properties of both metals and glasses. The term "metal" in metallic glass indicates that the material is made from metallic elements, while "glass" refers to its non-crystalline structure [1,2]. Theoretically, glass is a material that does not crystallize during the transition from liquid to solid. Typically, metal alloys crystallize upon cooling, with atoms arranged in an orderly manner to form crystalline metals like steel. However, rapid solidification prevents this, causing atoms to be "frozen" in a disordered state, thus creating a metallic glass [3]. On a microscopic level, metallic glass resembles a highly viscous liquid, also called "frozen melt". Common glassy materials like plastics, glass, rosin, paraffin, asphalt, and rubber share a disordered atomic or molecular arrangement. If the atomic arrangement of crystalline solids like steel is akin to a well-ordered parade, the arrangement in amorphous solids is like a bustling crowd on a busy street. Ordinary window glass is made from elements like silicon and oxygen; plastics are polymeric glasses, and metallic glasses are metal alloys composed of different metallic elements. For example, zirconium-based metallic glass is synthesized from zirconium, copper, aluminum, and titanium. Due to its unique structure, metallic glass exhibits excellent mechanical, physical, and chemical properties [4].

The discovery of glass dates back to around 3000 BC in ancient Babylon, indicating a long history. However, metallic glass appeared much later [5]. In 1960, Professor

Duwez from the California Institute of Technology reported the discovery of an amorphous structure in a rapidly solidified $Au_{75}Si_{25}$ alloy in a brief paper in *Nature* [6]. This sparked extensive research into amorphous alloys, leading to the commercial production of amorphous alloy ribbons and wires by Chen and colleagues in 1971 using melt spinning techniques (Hofmann, 2013). These alloys, known for their excellent soft magnetic properties, were widely used in transformer cores [7]. Early preparation techniques and limited alloy compositions restricted amorphous alloys to thin ribbons or wires, limiting their commercial application [8]. In the 1970s, Chen and colleagues produced palladium-copper-silicon amorphous alloy rods using suction casting methods, achieving millimeter-scale diameters and marking the earliest bulk amorphous alloys (Hofmann, 2013). Since the 1980s, various bulk amorphous alloy systems with dimensions exceeding 1 mm have been developed [9]. Notably, in 1993, Johnson and colleagues developed bulk zirconium-based amorphous alloys reaching centimeter scales, significantly promoting commercial applications and initiating a second research wave [10]. In the 21st century, the development of amorphous alloys accelerated, achieving significant milestones [11]. For instance, in 2005, academician Wang Weihua's team at the Institute of Physics, Chinese Academy of Sciences, developed metallic plastic, an amorphous alloy deformable like thermoplastic at around 90 °C but exhibiting metal characteristics at room temperature [11]. These properties enable researchers to imprint and mold amorphous alloys at suitable temperatures, fabricating nanoscale precision components unattainable with conventional crystalline alloys [12,13]. Despite only a 60-year history, amorphous alloys have seen significant theoretical and technological advancements [14], as summarized in Figure 1.

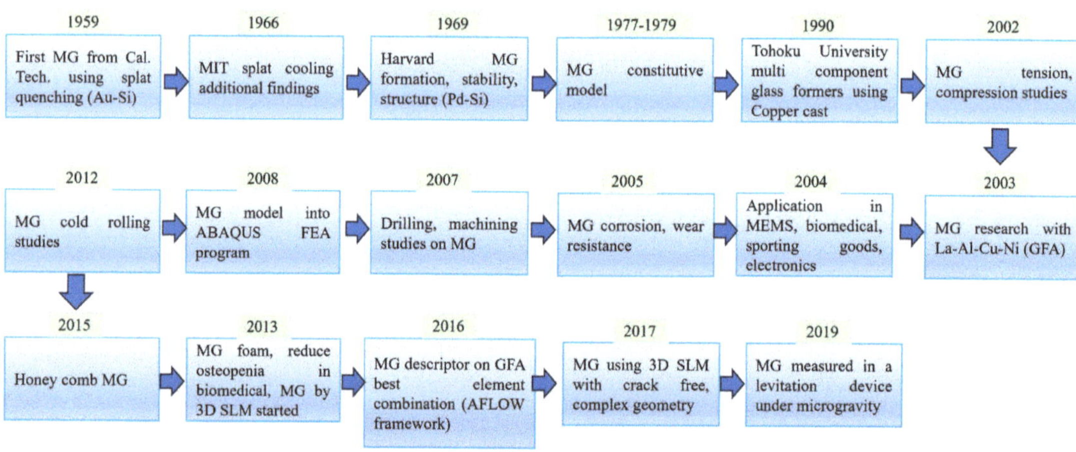

Figure 1. Historical development of amorphous alloys. Reprinted with permission from Ref. [15], Copyright 2021, copyright Springer Nature.

Amorphous alloys and their composites, known for their excellent properties (see Section 4), have broad applications in modern science and technology. These alloys are free from defects such as dislocations and grain boundaries, which are common in crystalline materials, and possess a super-high elastic limit. This makes them particularly well-suited for use in micro-electromechanical systems (MEMSs) [16]. For instance, the Pd-Cu-Si amorphous thin film micro-spring is utilized as a trigger in MEMS devices [16]. Additionally, the softening behavior observed in the supercooled liquid region of Zr-based amorphous alloys facilitates the easy manufacturing of complex-shaped precision components, such as those found in consumer electronics [17]. The abundance of active sites on the surfaces of amorphous alloys is leveraged in high-performance catalysts, exemplified by the Fe-Si-B-Nb thin film used for azo dye degradation [18]. Furthermore, the isotropic chemical properties of these alloys enable uniform chemical corrosion and dissolution, making them ideal for aerospace applications. A notable example is the Zr-Nb-Cu-Ni-Al amorphous alloy coating on the solar wind particle collection panel of NASA's Genesis spacecraft [19]. Finally, the ultra-high elastic limit of amorphous alloys enhances their suitability for sports equipment, like the Zr-based alloy baseball bat, which offers high energy transfer efficiency [20].

In recent years, researchers have discovered that amorphous alloys excel at suppressing harmonic vibrations compared to other metal materials. This ability effectively controls the reflection of noisy sound waves, resulting in a darker sound background, a stronger sound base, and excellent intrinsic low-frequency attenuation [20]. These properties also ensure good energy transmission, leading to fuller and more mellow sound, especially in the mid-frequency vocal range. Due to these outstanding acoustic properties, amorphous alloys have found significant applications in acoustics. For instance, Zr-based amorphous alloys have been used in headphone components by Sony (released in 2018) and for fixing pins in string instruments by Martin Guitar (released in 2015). Furthermore, amorphous alloys exhibit excellent forming properties in the supercooled liquid region, coupled with ultra-high strength, hardness, and wear resistance after cooling and solidifying. These characteristics make them ideal for creating high-precision, high-conformity, and strong creep-resistant components that are challenging to produce with conventional crystalline materials. Notable examples include the car door lock cover by Tesla (released in 2015) and the foldable phone hinge by Huawei (released in 2020) [20]. In addition to these applications, the largest use of amorphous alloys is in soft magnetic amorphous materials, such as transformer core materials. Transformer cores made from soft magnetic amorphous alloy materials achieve an energy conversion efficiency of 99.3%, compared with 97% for the best soft magnetic crystalline alloys [21]. With ongoing research, the potential application range of amorphous alloys is expected to expand significantly [22–24].

2. Formation and Production of Amorphous Alloys

2.1. Conditions for the Formation of Amorphous Alloys

a. Fundamental Principles of Amorphous Alloy Formation

(1) Cooling Rate and Glass-Forming Ability (GFA)

The formation of amorphous alloys primarily depends on the cooling rate. Different substances require vastly different cooling rates to form an amorphous state. For example, SiO_2 can form a glassy state at a cooling rate lower than 10^{-3} K/s, whereas many metals and alloy systems struggle to form an amorphous state even at a cooling rate of 10^{10} K/s. The ability to form an amorphous state is typically characterized by the critical cooling rate (RR_c) [25]. Studies have found that even for metallic systems, the difference in their forming ability can be more than 12 orders of magnitude. Generally, amorphous alloys with a critical size greater than 1 mm are called bulk amorphous alloys [26–28].

$$RR_C = \frac{dT}{dt\left(\frac{K}{s}\right)} = 10/D^2(\text{cm})$$

where D is the critical size.

The critical size refers to the maximum size of a bulk amorphous alloy that can be formed. Generally, amorphous alloys with a critical size greater than 1 mm are referred to as bulk amorphous alloys. The GFA of an alloy system can be measured by determining the maximum size of the bulk amorphous alloy that can be formed. Systems with good GFA can form amorphous alloys at relatively slow cooling rates and achieve larger critical sizes. To summarize, GFA measures the ease with which an alloy can be transformed into an amorphous state. It is influenced by the cooling rate required to prevent crystallization and the critical size of the alloy that can be produced in an amorphous form. Materials with high GFA can form amorphous structures at lower cooling rates and achieve larger critical sizes, making them suitable for forming bulk amorphous alloys.

(2) Thermodynamic Conditions

The formation of an amorphous state is closely related to the thermodynamic conditions of the substance. During the cooling process of a high-energy liquid, if the free energy is lower than that of the crystalline phase and the system can overcome the nucleation barrier, an amorphous state may form. This formation is a metastable phenomenon, where the free energy of the liquid or amorphous state is higher than that of the crystalline state within a certain temperature range, but due to kinetic reasons, the amorphous state persists [29].

(3) Kinetic Factors

The kinetic conditions for forming an amorphous state include avoiding crystal nucleation and growth during the cooling of the liquid metal. Turnbull's research indicates that liquid metals can be supercooled far below the equilibrium melting point without nucleation and growth [30]. In supercooled liquids, the nucleation rate and growth rate of crystals are crucial for forming an amorphous state. A key indicator of amorphous formation is sufficiently low nucleation and growth rates in the supercooled liquid, allowing it to maintain a disordered state [31].

b. Specific Conditions for Amorphous Alloy Formation

(1) Chemical Bonding Characteristics

The formation of amorphous alloys is closely related to the chemical bonding characteristics of the substance [32]. The type and strength of chemical bonds determine key parameters such as melting point, rheological properties, and viscosity coefficient. For example, substances with covalent and ionic bonds form network structures, which have high viscosity and favor the formation of amorphous states. Additionally, electronegativity affects the glass-forming ability; oxides composed of atoms with low electronegativity are difficult to form into an amorphous state.

(2) Atomic Size and Composition

Differences in atomic size and compositional complexity in alloy systems significantly affect the glass-forming ability. Egami's research suggests that when the atomic size difference between components exceeds 12%, it promotes a densely packed random atomic structure, enhancing the glass-forming ability [33]. Moreover, multi-component alloy systems with complex compositions have a high nucleation barrier, significantly increasing the tendency to form an amorphous state.

(3) Thermodynamic and Kinetic Stability

Methods to enhance the stability of amorphous alloys include increasing the crystallization temperature and the glass transition temperature, thereby achieving a wide supercooled liquid region. Alloy systems should possess high diffusion or nucleation barriers to maintain the stability of the amorphous state over a wide temperature range.

2.2. Rapid Solidification Techniques

Typical gas-phase processes include vapor deposition, such as physical vapor deposition (PVD) and chemical vapor deposition (CVD) [34]. The amorphous alloys prepared by vapor deposition are mostly thin film samples. A popular vapor deposition technique is ion beam sputtering deposition, a type of PVD [35]. This process uses a high-energy particle beam, such as an argon (Ar) ion beam, to bombard the target material's surface atoms, causing them to detach from the target. The detached atoms then deposit on a cold substrate to form an amorphous alloy film [26]. Due to the direct condensation from the vapor phase to the solid phase, this process achieves a cooling rate of up to 10^{13} Ks^{-1}. At such high cooling rates, many metals and their alloys can avoid crystallization and form amorphous alloys, and even pure metals can form pure amorphous states [35,36]. As an example, Wang et al. successfully prepared an amorphous iridium-nickel-tantalum (Ir-Ni-Ta) alloy film hydrogen evolution catalyst with high intrinsic activity using ion beam sputtering deposition [37]. The ion beam sputtering deposition technique they used is shown in Figure 2, which also illustrates the auxiliary ion source generally used for cleaning the deposited substrate.

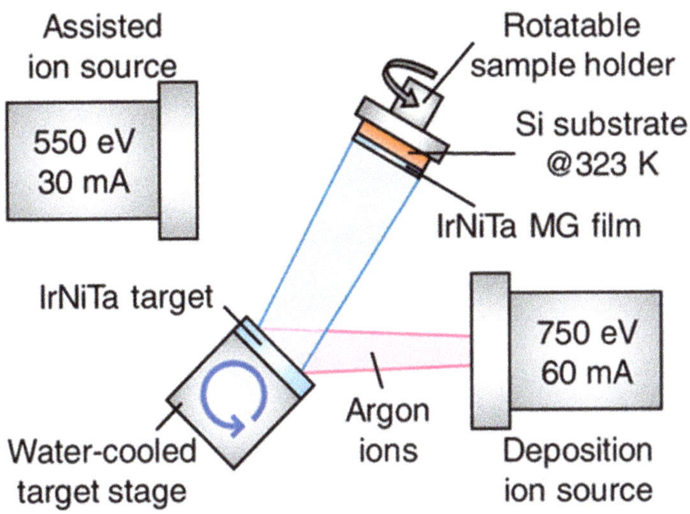

Figure 2. Schematic diagram of ion-beam sputtering deposition technology. Reprinted with permission from Ref. [37], Copyright 2020, copyright John Wiley and Sons.

There are various methods for preparing amorphous alloys through liquid processes, such as rapid solidification of alloy melts [38] and electrochemical deposition [39]. Common laboratory methods for preparing amorphous alloys through liquid processes include melt spinning, copper mold suction casting, and spray casting [40]. In the melt spinning process, the alloy is first melted into a liquid state through induction melting, then rapidly ejected onto a high-speed rotating copper wheel by applying pressure with an inert gas. This causes the melt to rapidly solidify and form an amorphous state before crystallization can occur [41]. In the spray casting process, the alloy is also first melted and then rapidly ejected into a cooled copper mold by applying pressure with an inert gas, forming an amorphous state. By altering the internal shape of the mold, the ability of the alloy system to form bulk amorphous alloys can be measured, i.e., the maximum size of the bulk amorphous alloy that can be formed [35]. In the copper mold suction casting process, the alloy ingot is melted into a liquid state through arc melting in a mold with small holes of different diameters at the bottom. Due to surface tension, the alloy melt typically does not automatically drop into the mold. A mechanical pump at the bottom of the mold generates a certain suction force to draw the alloy melt rapidly through the mold holes into the mold, where it is quickly solidified into an amorphous alloy [35].

The core of these liquid processes for preparing amorphous alloys is the cooling rate of the liquid melt. Typically, the melt spinning process can achieve a cooling rate on the order of 10^6 Ks^{-1}, while the cooling rate in the copper mold casting process is much lower, around the order of 10^3 Ks^{-1} [36]. Therefore, these methods have limited cooling rates and are restricted by the glass-forming ability of the alloy systems. Generally, only alloy systems with good glass-forming ability can be prepared into an amorphous state using these methods, such as certain Zr-based (e.g., $Zr_{65}Cu_{17.5}Ni_{10}Al_{7.5}$) and Pd-based (e.g., $Pd_{40}Cu_{30}Ni_{10}P_{20}$) alloy compositions [35]. For systems with exceptionally good glass-forming ability, such as the aforementioned Pd-based systems, even a simple water-quenching process of the melt can result in an amorphous state [42].

Electrochemical deposition offers another method for preparing amorphous metals or alloys via liquid processes. In the electrochemical deposition process, metals or alloys containing the desired elements are deposited through electrochemical reduction reactions from a precursor solution containing these elements. Theoretically, even pure metals with poor glass-forming ability can form an amorphous state using this method [43]. In fact, as early as 1950, Brenner et al. first used electrochemical deposition to prepare a series of phosphorus (P)-containing cobalt (Co) and nickel (Ni) alloys and reported that the X-ray diffraction pattern of a high-Ni-content P-containing alloy sample showed only a diffuse pattern, suggesting a possible amorphous phase [44]. Unfortunately, the authors did not provide the X-ray spectrum in their report, nor the exact composition of the alloy. Nevertheless, Brenner's work indicated the feasibility of preparing amorphous alloys through electrochemical deposition.

In recent years, Wang et al. reported a method for in situ electrochemical deposition of amorphous metals in button cells, which can even produce amorphous lithium (Li), a highly active metal [45]. A schematic diagram of this method is shown in Figure 3. Figure 3a illustrates a Li metal battery device, which includes a Cu current collector (for drawing out the negative current), a Li metal anode, electrolyte, separator, a Li transition metal oxide cathode, and an Al current collector (for drawing out the positive current). During the battery cycling process, lithium ions (Li^+) shuttle between the cathode and anode is driven by the internal electric field, undergoing electrochemical deposition on the electrodes. Figure 3b shows a schematic of Li^+ undergoing electrochemical deposition on the Li metal anode. With a reasonable device design and appropriate electrochemical deposition conditions, the amorphous metal Li can be obtained [45]. Although this in situ electrochemical deposition method often yields only a very small amount of amorphous metal in a single experiment, and the sample preparation and transfer process is rather cumbersome, it represents a significant advancement in the preparation of highly active metals in an amorphous state. Typical cooling rates for these techniques can range from a few 10^3 K/s for copper mold casting up to several thousand 10^6 K/s for melt spinning processes. Other techniques, such as electrochemical deposition and mechanical alloying, do not directly specify cooling rates but involve conditions that promote the formation of amorphous phases.

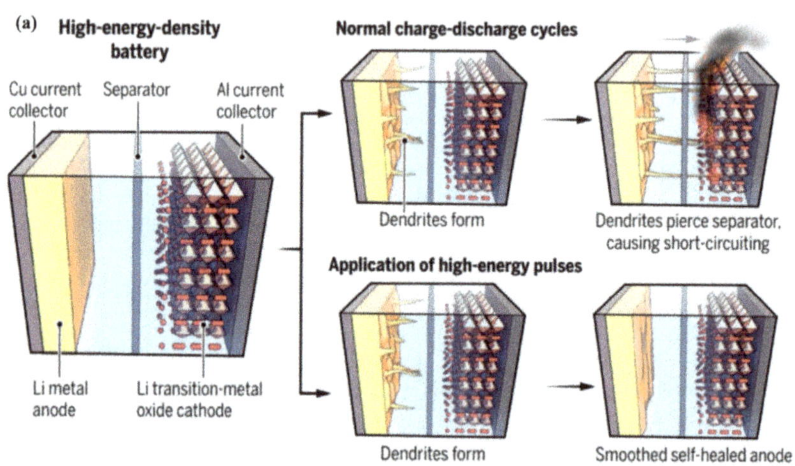

Figure 3. *Cont.*

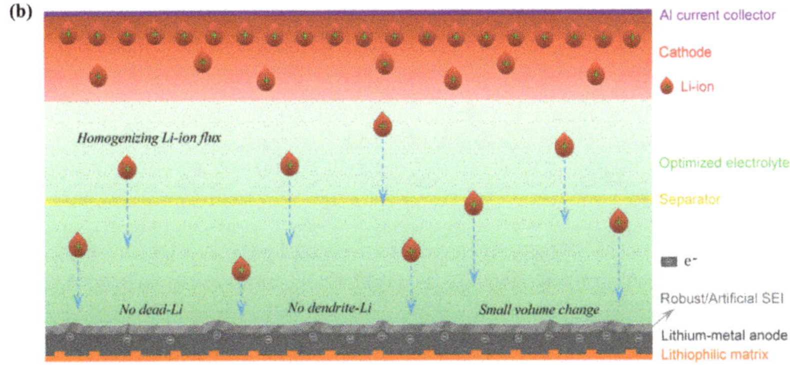

Figure 3. Schematic diagram of preparation of non-crystalline Li by in situ electrochemical deposition (**a**) Schematic diagram of a Li metal battery device (Reprinted with permission from Ref. [45], Copyright 2018, copyright The American Association for the Advancement of Science). (**b**) Schematic diagram of in situ electrochemical deposition of Li ions (Reprintedwith permission from Ref. [46], Copyright 2020, copyright ROYAL SOCIETY OF CHEMISTRY, ETC).

2.3. Other Production Methods

There are many methods for preparing amorphous alloys through solid-state processes, such as irradiation-induced amorphization [47], deformation-induced amorphization [48], and the rolling process [49]. However, the most commonly used method is mechanical alloying [1]. The basic principle of mechanical alloying is to induce severe deformation in the mixed metal powders through mechanical means. This process increases the defect content in the system and raises the Gibbs free energy, facilitating the transition from a stable state to a metastable state, such as the transformation from mixed elemental metal powders to amorphous alloy powders [50], as illustrated in Figure 4.

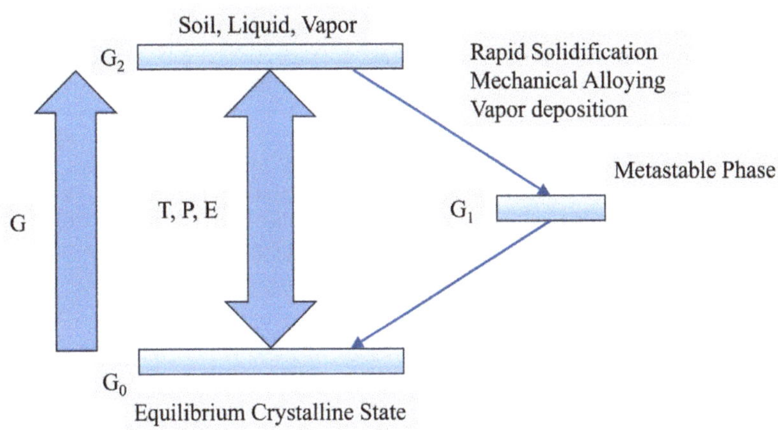

Figure 4. Basic principle of mechanical alloying process. Adapted with permission from Ref. [50], Copyright 2001, copyright Elsevier.

Thin film metallic glasses (TFMGs) are typically prepared using deposition techniques such as physical vapor deposition (PVD) and sputtering, which allow for high-quality amorphous films with uniform thickness [51]. These films exhibit superior mechanical properties compared to bulk metallic glasses, including higher strength and increased ductility. At the nanoscale, TFMGs can attain ceramic-like strengths and metal-like ductility simultaneously without the catastrophic failure associated with bulk materials. The reduction in size limits the formation of shear bands, promoting homogeneous deformation [52]. This section provides an overview of deposition techniques, unique film architectures, and their applications, focusing on their mechanical behavior. Pulsed Laser Deposition (PLD) is a powerful technique for synthesizing ultrafine nanolaminates (U-NLs) with precise control over local heterogeneities. PLD allows the creation of novel crystal/glass U-NLs, such as structures with ~4 nm thick crystalline Al layers separating 6 and 9 nm thick $Zr_{50}Cu_{50}$ glass nanolayers. This technique achieves a high density of sharp interfaces and significant chemical intermixing. Compact U-NLs, grown atom-by-atom, exhibit high mass density (~8.35 g/cm^3) and enhanced mechanical properties, including hardness and yield strength up to 9.3 and 3.6 GPa, respectively. In contrast, nanogranular U-NLs, formed via cluster-assembled growth, show slightly lower yield strength (3.4 GPa) but enhanced elastoplastic deformation (~6%) [53]. Multilayer films, consisting of alternating layers of different materials, provide unique opportunities to tailor the mechanical and functional properties of metallic glasses. These films can control shear band nucleation and propagation, resulting in improved strength and ductility. Nanogranular films featuring nanometer-sized grains dispersed in an amorphous matrix exhibit exceptional mechanical properties such as high hardness and wear resistance. These films are particularly promising for microelectronics and coatings applications. Their formation mechanisms and properties suggest significant potential in microelectronics, particularly in MEMS and NEMS devices, where high-performance electronic components are essential. In the coatings industry, TFMGs offer excellent corrosion and wear resistance, making them ideal for protective coatings in aerospace and automotive applications [54]. The robust properties of these films are crucial for industries requiring durable and long-lasting materials. The combination of high hardness, yield strength, and elastoplastic deformation makes TFMGs suitable for environments where mechanical stress and exposure to harsh conditions are common. In conclusion, the deposition techniques and unique architectures of TFMGs enable the development of materials with superior mechanical properties and a wide range of applications. The advancements in PLD and the exploration of compact versus nanogranular films highlight the versatility and potential of TFMGs in various high-tech industries.

It should be noted that the product obtained from the mechanical alloying process is usually amorphous alloy powder, which often requires further processing to achieve the desired final product. The most common method for achieving mechanical alloying is high-energy ball milling. In this method, pure metal elemental powders or pre-alloyed powders in specific proportions are mixed with quenched steel balls or hard alloy balls in a certain ratio and placed in a ball mill jar. The jar can be filled with a specific atmosphere or vacuum. The prepared jar is then mounted on a ball mill, and the milling parameters, such as rotation speed and milling time, are set according to the desired outcome. Ultimately, the required amorphous alloy powder can be obtained [50]. Figure 5 shows a planetary high-energy ball mill and its ball mill jar.

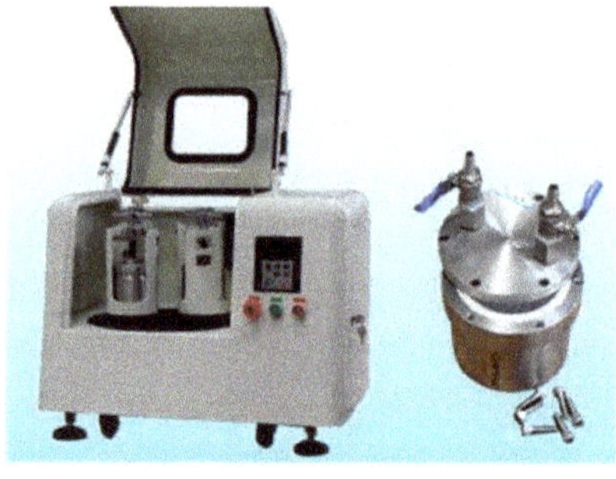

Figure 5. A planetary high-energy ball mill (**left**) and its milling jar (**right**).

3. Structure and Properties

3.1. Description of the Atomic Structure of Amorphous Alloys

The earliest structural models of amorphous alloys may originate from studies on the structure of liquids. Specifically, early explorations of mathematical and physical models of liquid structures by Bernal and colleagues at the University of Cambridge in the 1960s laid the foundation for qualitative structural models of amorphous alloys. A key point in this was the introduction of the concept of random close packing [55]. Because glass is also considered a "frozen liquid" [56], amorphous alloys, also known as glassy alloys or metallic glasses, naturally lend themselves to the application of the random close packing concept in their structural modeling. This involves modeling the structure of amorphous alloys using the random close packing of equal-sized hard spheres, which may be among the earliest studies on the structural models of amorphous alloys [53].

However, this model has several deficiencies. One significant issue is that it does not address the structure of amorphous alloys containing components of different sizes and concentrations. Additionally, the introduction of components of varying sizes and concentrations makes the model analysis extremely difficult. Furthermore, the nearest neighbor coordination in this model is similar to the crystalline phase, seemingly not meeting the requirements of randomness and being difficult to extend beyond the nearest neighbor coordination shell to obtain the medium-range order discovered later in amorphous alloys [57].

From 2004 onwards, Miracle et al. further proposed and gradually refined the cluster-dense packing model for amorphous alloys [57–60], as shown in Figure 6. In this model, the authors proposed that the basic units for constructing the structure of amorphous alloys are not the atoms of the components themselves but clusters centered around solute atoms that can pack effectively, such as FCC and HCP clusters, which are favorable because they can most efficiently fill space [60]. These clusters, as basic units, can be idealized as spheres packed in three-dimensional space, filling the space [51]. The order of the solute forming the clusters does not extend beyond even a few cluster distances due to internal strain, thus not forming long-range order or orientational order [61]. Adjacent clusters share solvent atoms in a vertex-sharing, edge-sharing, or face-sharing manner, causing the clusters to overlap in the nearest neighbor coordination shell, with solvent atoms randomly occupying positions. From a topological perspective, regardless of the number of elements in the alloy, there are only three types of solutes, each with a specific size relative to the solvent, to produce effective atomic packing [57].

The schematic of this model is shown in Figure 6a,b. In the left part of Figure 6a, Ω represents solvent atoms (within the first shell), and the three types of solutes are α as the solute forming the primary cluster or the cluster center, β as the solute occupying the octahedral interstices of the cluster, and y as the solute occupying the tetrahedral interstices of the cluster (not illustrated in the Figure) [57]. From the perspective of structural sites, combining Ω, the cluster-dense packing model suggests that there are no more than four different types of sites (Ω, α, β, y) in the structure of amorphous alloys. The occupation of these sites can provide information on the minimum solute concentration for forming amorphous alloys [60]. If some sites are unoccupied, this corresponds to vacancies in the structure; if the solvent occupies solute sites, such as Ω occupying the α site, it forms antisite defects. The right part of Figure 6a shows the three-dimensional packing of solvents and various clusters. Here, α solutes are wrapped by pink solvents Ω, β solutes are purple spheres, and y solutes are orange spheres. Figure 6b shows a cluster organized in a hexagonal arrangement with vertex-sharing and face-sharing connections (bottom right inset). This packing method imagines the cluster as a sphere packing (spheroidal packing) to fill space. Since clusters are not true spheres, the gaps and contact modes between clusters (as shown in the bottom right inset) vary, causing distortions and deviations from the strict lattice form of HCP packing, resulting in an amorphous structure [21].

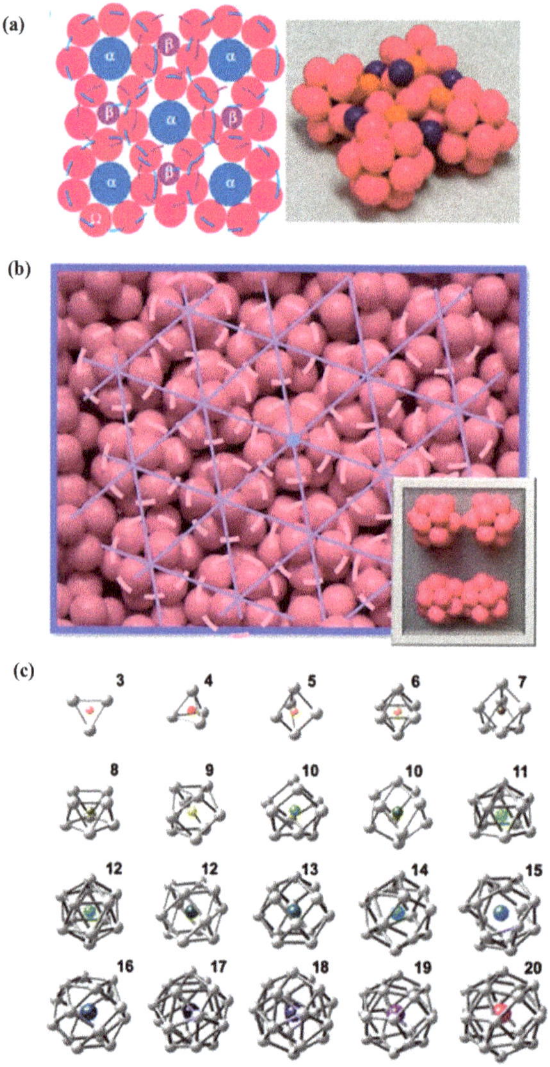

Figure 6. The cluster dense packing structure model of amorphous alloys. (**a**) Basic structural unit of the cluster dense packing model (Reprinted with permission from Ref. [57], Copyright 2004, copyright Springer Nature). (**b**) HCP organization form of the clusters, and (**c**) some clusters that can effectively pack, with numbers indicating coordination numbers (Reprinted with permission from Ref. [21], Copyright 2012, copyright Springer Nature).

The cluster-dense packing model has achieved some theoretical and experimental successes. First, the structure of amorphous alloys can be constructed purely from topology (relative sizes and number of components), involving no more than four topologically different sites: Ω, α, β, and y or one solvent Ω and three solutes α, β, and y (only three topologically different ones) [57]. Depending on the packing of clusters, defects may appear in the structure, such as vacancies like β, y vacancies, and antisite defects like Ω occupying the α site. This construction links the static structural model to some dynamic processes in amorphous alloys, such as diffusion and relaxation. Second, early structural models of amorphous alloys could not explain the physical origin of short-range order and medium-range order. The cluster-dense packing model provides a simple and intuitive physical origin for these orders: the clusters as basic structural units are themselves short-range ordered, consistent with the short-range order in crystalline alloys of the same composition as amorphous alloys, such as FCC and HCP clusters. Short-range order is highly sensitive to topology, electronic structure, and even minor compositional changes, controlling the formation and stability of amorphous alloys, which tend to crystallize into crystalline phases with the same local order (including short-range and medium-range order) as corresponding crystalline alloys [57]. Medium-range order is formed by the spatial organization of overlapping clusters, with the medium-range order scale predicted by the cluster-dense packing model being about 1 nm [58,61]. Third, experimental characterization has found that the density of amorphous alloys is often only about 0.5% lower than that of crystalline alloys of the same composition [21], indicating a very high atomic packing efficiency in amorphous alloys. The density of amorphous alloys obtained from the cluster-dense packing model matches well with experimentally measured densities. Furthermore, the compositions of many early binary amorphous alloys and the diffraction experimental results of the short-range and medium-range order they contain also match the predictions of the cluster-dense packing model [62]. However, the cluster-dense packing model is not yet perfect. As pointed out by the original authors of the model, it is not entirely predictive and still requires quantitative descriptions of chemical interactions to determine how topologically equivalent but chemically significantly different solutes enhance or inhibit the stability of amorphous alloys [57]. Additionally, the packing fractions given by the model are sometimes unrealistically high, which is another shortcoming [21].

This section mainly reviews two typical structural models proposed during the study of amorphous alloy structures. In fact, researchers have proposed various structural models for amorphous alloys, but they generally revolve around sphere packing (including later-developed soft sphere packing) and cluster packing [63]. Currently, there is no consensus on the structure of amorphous alloys, and many different structural models exist for different amorphous alloy systems. These models can be effective in solving specific system problems but still have many shortcomings in quantitative descriptions [63]. Solving the structural issues of amorphous alloys remains a challenging and long-term task.

3.2. X-ray Diffraction and Transmission Electron Microscopy Analysis

X-ray diffraction (XRD) techniques and their derivative methods have been extensively utilized to investigate various aspects of amorphous alloys. These techniques range from confirming the amorphous state to analyzing the microstructure, such as short-range order, medium-range order, and coordination number, as shown in Figure 7. Figure 7a presents the conventional XRD spectrum of the Ir-Ni-Ta amorphous alloy system, which is frequently used in contemporary laboratories to preliminarily confirm the amorphous state of samples. Figure 7 confirms that a series of Ir-Ni-Ta amorphous alloys are indeed amorphous [64]. Moreover, conventional XRD spectra can also reflect specific local structural information. For instance, Figure 7b shows the conventional XRD spectra of a set of $Al_{(90-x)}Ni_{10}Ce_x$ (x = 3, 5, 6, 8) amorphous alloys. Besides the main peak, a distinct prepack appears on the low-angle side of the XRD spectra. The original authors attributed this to the medium-range ordered cluster structure related to cerium (Ce) in these amorphous alloys [65]. It is worth noting that the original image of this Figure lacks units on the horizontal axis 2θ, which might have been omitted in the original text. However, this omission does not affect the interpretation of Figure 7.

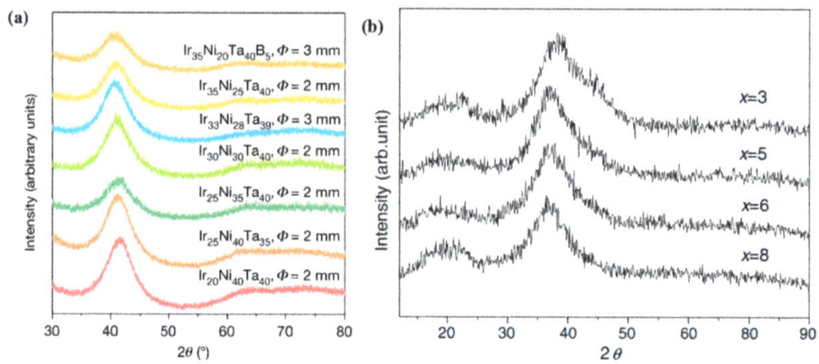

Figure 7. Various X-ray diffraction techniques are used to characterize the structure of amorphous alloys. (**a**) Conventional XRD spectrum of Ir-Ni-Ta amorphous alloy (Reprinted with permission from Ref. [64], Copyright 2019, copyright Springer Nature). (**b**) Conventional XRD spectrum of Al(en)NinoCes (x = 3, 5, 6, 8) amorphous alloy with a prepack (Reprinted with permission from Ref. [65], Copyright 2004, copyright Elsevier).

With the continuous development of synchrotron radiation technology, various X-ray techniques based on synchrotron radiation can also effectively characterize the structure of amorphous alloys [66–68]. In addition to the more common high-energy synchrotron XRD, techniques such as Extended X-ray Absorption Fine Structure (EXAFS) [69] and X-ray Cross-Correlation Analysis (XCCA) [70] are also employed, as shown in Figure 8. Figure 8 displays the EXAFS spectra at the Cu-K and Zr-K absorption edges (left) and their corresponding Fourier transform spectra (right) for $Cu_{50}Zr_{50}$, $Cu_{45}Zr_{45}Ag_{10}$, and $Cu_{40}Zr_{40}Ag_{20}$ amorphous alloys [69]. These spectra reveal the variations in the nearest-neighbor coordination of Cu and Zr in three different compositions of amorphous alloys. For example, the Cu-K edge spectra of $Cu_{50}Zr_{50}$ and $Cu_{45}Zr_{45}Ag_{10}$ in the upper section almost overlap, indicating that the addition of 10 at.% silver (Ag) hardly affects the nearest-neighbor coordination of the Cu shell [69]. However, the Cu-K edge absorption spectrum of $Cu_{40}Zr_{40}Ag_{20}$ differs significantly, showing splitting, which the authors attribute to possible phase separation in the sample [69]. The Zr-K EXAFS spectra in the lower section of Figure 8 show significant changes with the addition of Ag, especially on the low wavenumber side. The Fourier transform spectra on the right indicate that this change might be related to the inconsistency in peak intensity of the corresponding Zr-Zr atomic pairs. Since peak intensity is related to the number of neighboring atoms, and considering that the Cu-K EXAFS spectra of $Cu_{50}Zr_{50}$ and $Cu_{45}Zr_{45}Ag_{10}$ in the upper section show almost no change, the authors inferred that the addition of Ag selectively replaced some Zr-Zr atomic pairs with Zr-Ag atomic pairs [69]. Wochner P shows a schematic diagram of the XCCA technique apparatus. An incident coherent X-ray beam is collimated and then directed onto the sample, such as the colloidal sample shown in the figure, resulting in a two-dimensional diffraction pattern [70]. Through quantitative analysis (the diagram illustrates the construction of the cross-correlation analysis function), information about the local symmetry within the amorphous structure can be obtained. The diagram demonstrates the local five-fold symmetry (LFS) of PMMA (polymethyl methacrylate, also known as acrylic glass) spheres [70].

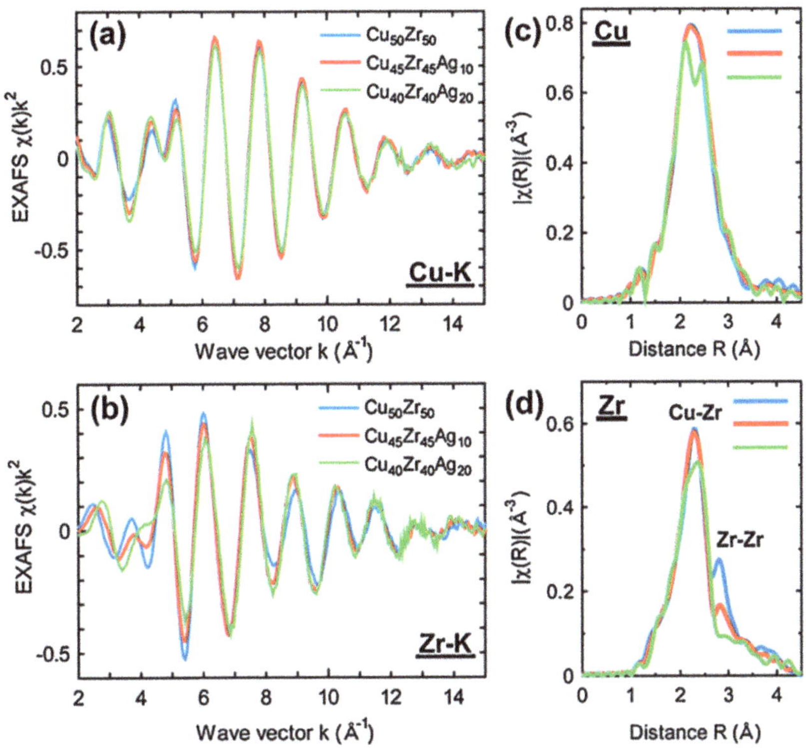

Figure 8. Synchrotron-based X-ray techniques are used to characterize the local structure of amorphous alloys. Reprinted with permission from Ref. [69], Copyright 2009, copyright American Physical Society.

Despite the significant convenience and rich structural information provided by X-ray diffraction (XRD) techniques for characterizing amorphous alloys, certain limitations necessitate the supplementation of other characterization techniques. X-ray-based characterization techniques typically target the overall amorphous alloy sample and often record the intensity information of X-rays scattered by the sample during the experiment, lacking phase information. The results obtained from analyzing intensity information generally reflect the structural information of the entire sample as a statistical average [71]. Consequently, these techniques cannot provide more detailed atomic-level structures and local environments, such as the short-range and medium-range order in amorphous alloys.

Furthermore, the structural information extracted by characterization methods based on X-ray diffraction techniques, such as RDF (which reflects the correlation between two atoms or atomic pairs rather than multi-body correlations), often only indirectly reflects the statistical distribution of atomic distances and atomic coordination in amorphous alloys [72]. These methods cannot provide direct visual structural images. While X-ray-based characterization techniques can produce reconstructed images, their resolution is relatively low [73]. For instance, computer-assisted tomography (CT) can reconstruct microstructural images of materials based on X-ray tomography. However, the spatial resolution of CT is limited and cannot reveal the precise structural or chemical composition information of different regions. Even though the latest third-generation synchrotron radiation sources can produce intense and bright X-rays, allowing for relatively short acquisition times and dynamic CT, the resolution of the reconstructed images is constrained by the physical size of the X-ray beam used, typically in the micrometer range [74]. Advanced nano-CT (with

X-ray beam sizes on the nanoscale) has a resolution of only a few tens of nanometers [73,75], which far exceeds the scale of characteristic local structures in amorphous alloys, such as medium-range order, which is approximately 5–20 Å [72]. This resolution is much larger than the scale of local structures in amorphous alloys, and even tiny nanocrystals cannot be detected by nano-CT.

Characterization techniques based on electron diffraction provide a highly resolved method for imaging typical local structures in amorphous alloys. Nowadays, advanced transmission electron microscopy (TEM) can achieve sub-angstrom resolution. TEM with spherical aberration correction can easily achieve atomic-level resolution, allowing the observation of phenomena such as the formation of metallic bonds between two metal atoms [76] and interstitial atoms in solid solutions [77]. TEM used for observing the structure of amorphous alloys can provide direct visual images, directly imaging various local orders [78], structural heterogeneities [79], and compositional fluctuations [80] in amorphous alloys.

In recent years, significant advances have been made in the study of the atomic-level structure of amorphous alloys, driven by advanced TEM characterization techniques. Around 2011, Chen and colleagues developed Aemi Beam Electron Diffraction (ABED) techniques, which are capable of directly observing short-range order, such as icosahedral order, in amorphous alloys, as illustrated in Figure 9 [81,82]. Figure 9a shows the ABED experimental setup, with a coherent electron beam diameter of 3.6 Å. This probe size is close to the characteristic local structures in amorphous alloys (typically, the short-range order in amorphous alloys is believed to be 2–5 Å, medium-range order is 5–20 Å, and the long-range order is over 20 Å [10]), enabling the direct detection of these local structures. Figure 9b presents diffraction patterns in the 2-, 3-, and 5-fold symmetry axis directions of an ideal icosahedron from molecular dynamics simulations. The original literature notes that the experimentally observed diffraction patterns in these directions retained only part of the corresponding axial symmetry of the ideal icosahedron. The discrepancy between experimental and simulated results is attributed to some distortion in the observed icosahedrons. When using distorted icosahedrons for simulation, the results closely match the experimental observations, as shown in Figure 9c [81]. Breakthroughs in advanced aberration-corrected electron microscopy, data acquisition, three-dimensional image reconstruction, and atomic tracking algorithms have made atomic electron tomography (AET) a more powerful tool for characterizing the structure of materials.

By collecting projection images of the sample from multiple directions and combining them with computer-aided reconstruction [83], as shown in Figure 10, AET can determine the three-dimensional grain boundaries of crystalline materials, the three-dimensional structure of dislocation core, and track the coordinates of individual atoms within the sample [83].

Atomic Electron Tomography (AET) has significantly advanced the determination of the three-dimensional atomic structure of amorphous alloys by enabling the tracking of individual atomic coordinates. In 2021–2022, researchers such as Yang [84] and Yuan [85] utilized this technology to map the three-dimensional atomic structures of multi-component (Co, Ni, Ru, Rh, Pd, Ag, Ir, Pt) amorphous nanoparticles, pure Ta amorphous films, and pure Pd amorphous nanoparticles, as illustrated in Figure 11. Figure 11a,b showcase the AET three-dimensional reconstructed images of the multi-component amorphous nanoparticles and the four typical medium-range ordered atomic packing configurations within them. In Figure 11a, due to the limitations of current AET technology, only three types of atoms in the multi-component amorphous nanoparticles can be distinguished: Type1 (Co or Ni), Type2 (Ru or Rh or Pd or Ag), and Type3 (Ir or Pt). Consequently, the reconstructed image displays only these three types of atomic coordinates [84]. Figure 11b illustrates four medium-range ordered structures found in the amorphous nanoparticles: FCC-like, HCP-like, BCC-like, and SC-like atomic packing states, containing 22, 14, 11, and 23 solute centers (large red spheres), respectively. These single solute-centered clusters (dashed circles) are randomly distributed. On the right side of these medium-range ordered structures, the quasi-lattice arrangements of solute-centered atoms along specific axial directions are shown after the removal of solvent atoms, though they significantly deviate from ideal lattice structures [84]. Figure 11c displays the AET three-dimensional reconstruction results of pure Ta amorphous films and Pd amorphous nanoparticles. The gray areas on the surface represent some crystal nuclei. Figure 11d depicts the multiple tetrahedral packing of atoms in these two pure amorphous substances, with the tetrahedral centers indicated by small black spheres. The connections between tetrahedral centers form triangular, quadrilateral, pentagonal, and hexagonal bipyramidal configurations [85], with brown spheres representing capping atoms.

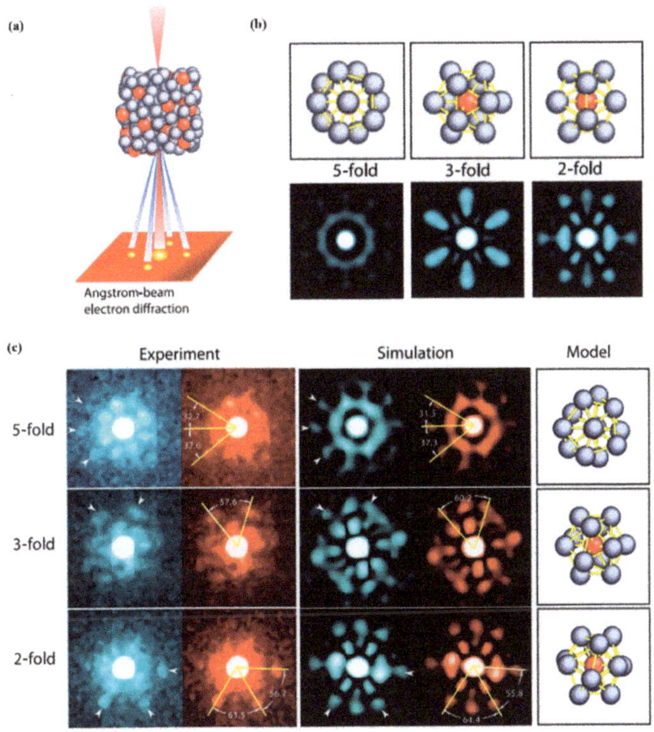

Figure 9. The ABED results of icosahedral short-range order in $Zr_{80}Pt_{20}$ amorphous alloy. (**a**) Schematic diagram of the ABED experiment (Reprinted with permission from Ref. [81], Copyright 2013, copyright Science). (**b**) Ideal icosahedral 5-, 3-, and 2-fold symmetry axis diffraction pattern from molecular dynamics simulation (Reprinted with permission from Ref. [81], Copyright 2013, copyright Science). (**c**) Comparison of experimentally observed and molecular dynamics simulated distorted icosahedral 5-, 3-, and 2-fold symmetry axis diffraction patterns (Reprinted with permission from Ref. [81], Copyright 2013, copyright Science).

These studies experimentally confirm the pervasive presence of short-range order in amorphous alloys and pure substances, as well as the formation of medium-range order through the interconnection of these short-range ordered superclusters [84]. The experimental results also partially validate the dense cluster packing model of the amorphous alloy structure proposed by Miracle and others [59]. Additionally, the AET reconstructed atomic structures highlight the inherent compositional and structural heterogeneity of amorphous alloys.

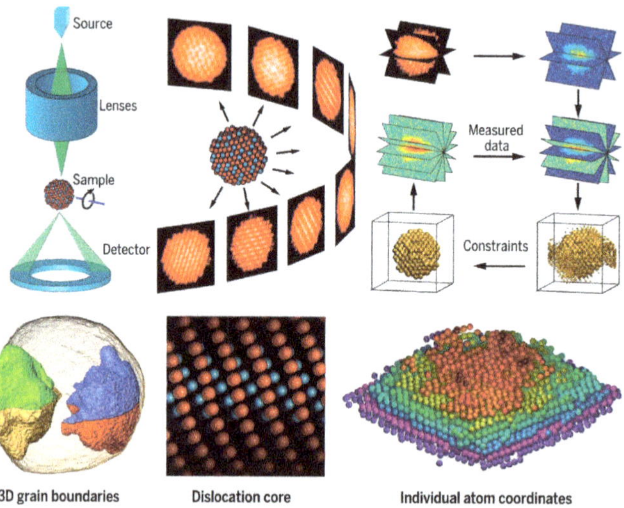

Figure 10. Atomic-level electron tomography (AET) technique and its applications. Reprinted with permission from Ref. [83], Copyright 2016, copyright Science.

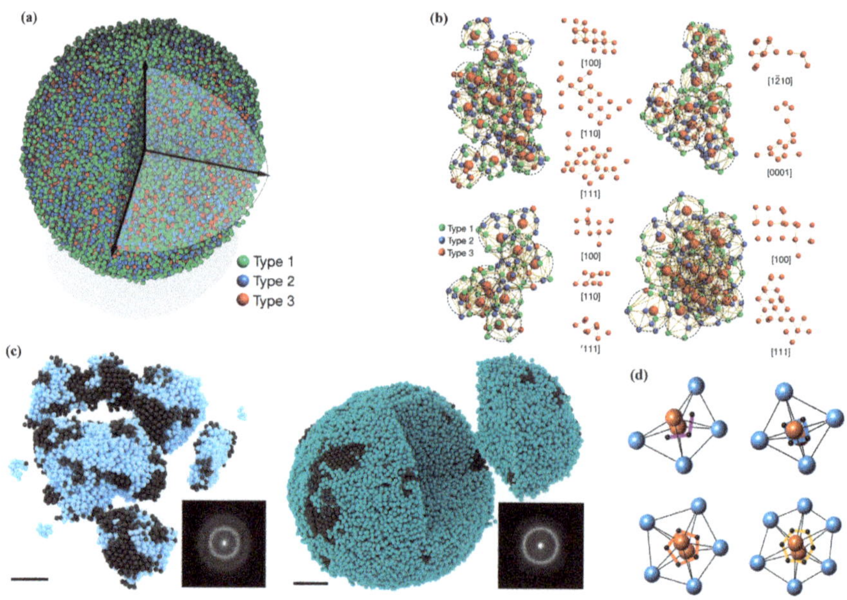

Figure 11. Atomic structure reconstructed by atomic electron tomography (AET) technology of multi-component (Co, Ni, Ru, Rh, Pd, Ag, Ir, Pt) amorphous nanoparticles and elemental Ta amorphous thin films, elemental Pd amorphous nanoparticles. (**a**,**b**) Three-dimensional reconstruction of atomic structure of multi-component amorphous nanoparticles and medium-ordered atomic accumulation therein (Reprinted with permission from Ref. [84], Copyright 2021, copyright Springer Nature). (**c**) Three-dimensional reconstruction of atomic structure of elemental Ta (left) amorphous thin films and Pd (right) amorphous nanoparticles (Reprinted with permission from Ref. [85], Copyright 2022, copyright Springer Nature). (**d**) Polytetrahedral packing found in elemental amorphous Ta films and Pd nanoparticles (Reprinted with permission from Ref. [85], Copyright 2022, copyright Springer Nature).

4. Physical and Chemical Properties

4.1. Mechanical Properties

Mechanical properties are key indicators for structural materials, and amorphous alloys have set numerous records for the mechanical properties of metal materials, such as the highest elastic limit [19], highest fracture strength [86], and highest fracture toughness [87]. Amorphous-nanocrystalline composites can even achieve superior mechanical properties [88], with some summarized in Figure 12. Due to the structural defects in amorphous alloys being confined to a few atomic dimensions, they do not exhibit the typical slip along specific planes seen in crystalline materials under external forces. This results in unique deformation behavior and mechanical property characteristics. High strength is the most notable mechanical property of amorphous alloys. For example, the fracture strength of bulk amorphous alloys can reach 1023 MPa for Mg-based [89], 2083 MPa for Ti-based [90], 2477 MPa for Zr-based [91], 3542 MPa for Ni-based [92], and 3280 MPa for Fe-based alloys [93]. Notably, the fracture strength of Co-based amorphous alloys is even more remarkable, with reports indicating that $Co_{43}Fe_{20}Ta_{5.5}B_{31.5}$ can reach 5185 MPa [94], whereas the strength of crystalline materials of the same alloy system is only a fraction or even a tenth of that of amorphous alloys.

Similarly, the resistance of amorphous alloys to indentation by hard objects is much higher than that of their crystalline counterparts. For instance, the Vickers hardness of $Fe_{41}Co_7Cr_{15}Mo_{14}C_{15}B_6Y_2$ [19] can reach 12.53 GPa, while the hardness of ordinary crystalline iron alloys is only about 0.2 MPa. The elastic modulus is an important physical quantity that describes the elasticity of a material, and it varies with the system and composition. Although the elastic modulus of amorphous alloys is lower than that of their crystalline counterparts, their elastic strain is around 2%, more than four times higher than that of high-carbon spring steel. Additionally, due to their high elastic limit, bulk amorphous alloys have a very high elastic specific energy.

Fracture toughness reflects the ability of a material to resist crack instability and propagation, which is crucial for the strength design of components. With the continuous optimization of the composition of amorphous alloys, their fracture toughness has significantly improved. For example, the fracture toughness of Zr-based [95] amorphous alloys reached 86 $MPa \cdot m^{1/2}$. Early amorphous alloys exhibited inhomogeneous strain under external forces, making fatigue cracks prone to nucleation under fatigue stress, resulting in lower fatigue life compared with traditional crystalline materials. This is mainly reflected in the differences in fatigue stress and cycle numbers. However, with continuous efforts by researchers, the fatigue properties of some bulk amorphous alloys have reached levels similar to traditional crystalline materials. For example, Gilbert [96] studied the crack propagation behavior of Zr-Ti-Cu-Ni-Be amorphous alloys and found that the stress range applied in this system determined the crack propagation rate. The load ratio had a similar effect to crack closure, and the fatigue fracture surface exhibited fatigue cracks similar to those in ductile metals.

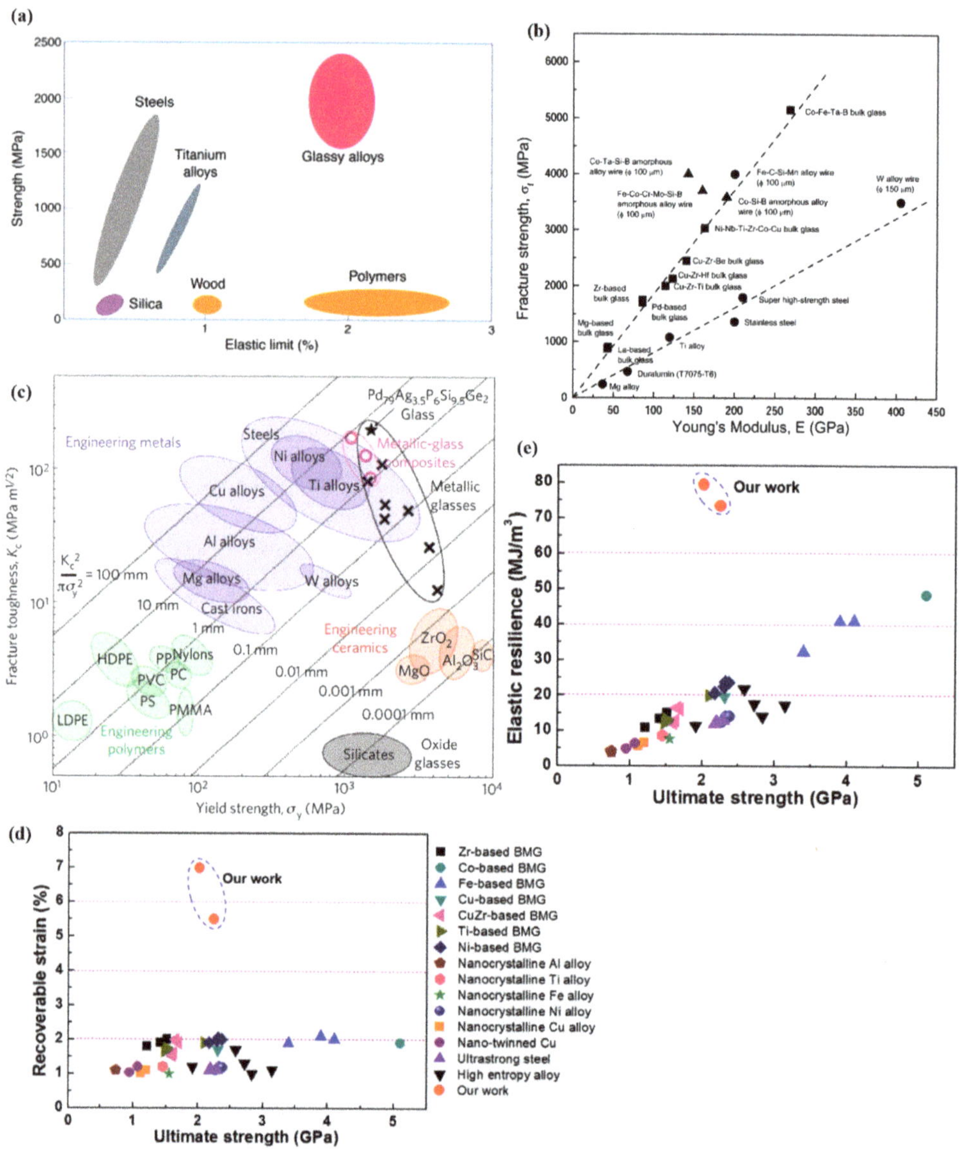

Figure 12. Mechanical properties of amorphous alloys. (**a**) Elastic limit-strength diagram of amorphous alloys (Adapted with permission from Ref. [19], Copyright 2004, copyright Elsevier). (**b**) The highest fracture strength of Co-Fe-Ta-B amorphous alloy (Adapted with permission from Ref. [86], Copyright 2004, copyright Elsevier). (**c**) The highest fracture toughness of $Pd_{79}Ag_{3.5}P_6Si_{9.5}Ge_2$ amorphous alloy (Adapted with permission from Ref. [87], Copyright 2011, copyright Springer Nature). (**d**,**e**) TiNi-based amorphous nanocrystalline composites can achieve higher elastic specific work and elastic limits (Adapted with permission from Ref. [88], Copyright 2020, copyright Elsevier).

4.2. Thermal Properties

The temperature range between the glass transition temperature (Tg) and the crystallization temperature (T_x) is referred to as the "supercooled liquid region". In this region, amorphous alloys exhibit high viscosity and varying degrees of superplastic deformation behavior [97]. These characteristics enable the forging and extrusion of amorphous alloys to produce micro gears and precision instrument components. One of the most prominent features of amorphous alloys is their excellent magnetic properties. Due to the absence of grain boundaries, there are no obstacles, such as precipitated phase ions, to pin the domain walls, making amorphous alloys with soft magnetic properties easy to magnetize and extremely low coercivity. Additionally, the higher electrical resistivity in amorphous alloys can significantly reduce eddy current losses associated with changes in the direction of magnetic domains. In terms of electrical properties, the long-range disorder of atomic arrangement in amorphous alloys enhances their ability to scatter electrons. At room temperature, their resistivity is generally 100–300 $\mu\Omega\cdot$cm, which is 2–3 times that of traditional crystalline alloys. Furthermore, the resistivity of amorphous alloys is less affected by temperature, resulting in a smaller temperature coefficient of resistivity compared to crystalline alloys.

4.3. Chemical Properties

Amorphous alloys exhibit high corrosion resistance due to the absence of defects in their structure and lack of compositional segregation. This results in a uniform atomic structure and chemical composition, leading to widespread application. Studies have shown significant hydrogen absorption capabilities in amorphous alloys, especially Mg-based alloys with a hydrogen adsorption ratio close to 100% [98]. When used as electrode materials, Pd-based amorphous alloys demonstrate a strong ability to generate chlorine gas, with minimal material consumption after 500 cycles, less than one-tenth of that seen in general electrode materials, which degrade significantly after 20 or fewer cycles [99]. The passivation film formation rate in amorphous alloys is faster and more uniform compared with crystalline alloys. Gao [100] studied Ca-Mg-Zn series amorphous alloys, finding that their bioabsorbable element composition makes them suitable for use as absorbable materials in orthopedic surgery. Guo [101] investigated the corrosion resistance of $Zr_{62.3}Cu_{22.5}Fe_{4.9}Al_{6.8}Ag_{3.5}$ amorphous alloy in artificial seawater using electrochemical methods, concluding that Zr-based amorphous alloys exhibit lower corrosion current density and potential.

5. Applications

5.1. Aerospace Industry

Amorphous alloys will have important applications in high-tech fields, which can mitigate their high cost while fully utilizing their unique properties. For example, amorphous alloys are considered essential candidate materials in the aerospace industry. The elastic deformation behavior of amorphous alloys is particularly significant for key structural materials that rely on elastic energy deployment. The elastic deformation of amorphous alloys can reach up to 2%, and their current maximum elastic limit exceeds 5000 MPa. Among lightweight amorphous alloys, titanium-based variants have an elastic limit exceeding 2000 MPa, which is unattainable by conventional crystalline and polymer materials. Recently discovered single-phase amorphous alloys with tensile strengths greater than 1.5 GPa and fracture toughnesses up to 200 MPa$\sqrt{m}$ represent the highest fracture toughness among materials, achieving a perfect combination of high strength and high toughness. Therefore, amorphous alloys can meet the stringent performance requirements of large deployable structures in spacecraft. NASA collaborated with Professor Johnson's research group at the California Institute of Technology shortly after the discovery of bulk amorphous alloys to develop high-hardness, high-specific-strength amorphous alloy foam materials. They also utilized the chemical homogeneity of amorphous alloys for solar wind collector materials [102].

5.2. Biomedical Engineering

Another important characteristic of amorphous alloys is their biocompatibility, degradability (such as in Ca- and Mg-based amorphous alloys), and non-allergenicity. These properties make them suitable for medical applications, including repairing implants and manufacturing surgical devices like surgical knives, artificial bones, biosensing materials for electromagnetic stimulation within the body, and artificial teeth [103]. Biodegradable biomaterials for implantation in the body are advantageous because they can avoid the need for secondary surgeries to remove the implant and prevent biological rejection issues associated with permanent implants. Mg-based amorphous alloys, due to their degradability, high strength, and elastic modulus close to that of bone, have the potential to become the next generation of materials for internal scaffolds. Thus, Mg-based amorphous alloys hold great promise in the field of biodegradable biomaterials.

5.3. Electronics and Energy Technology

The most mature and widely used application of amorphous alloys is in the field of amorphous magnetism [103]. Fe-, Ni-, and Co-based amorphous alloy ribbons have found extensive applications due to their excellent soft magnetic properties. These ribbons have become the ideal core materials for various transformers, inductors, sensors, magnetic shielding materials, and radio frequency identifiers. They are now indispensable fundamental materials in the fields of power, power electronics, and electronic information, with their manufacturing technology being quite mature.

Recently, Tohoku University in Japan developed various ferromagnetic bulk amorphous alloys and prepared toroidal magnetic cores using copper mold casting. These bulk amorphous alloys exhibit high magnetic saturation strength, high permeability, low coercivity, and low saturation magnetostriction, making their soft magnetic properties far superior to those of traditional silicon steel sheets and conventional crystalline magnetic materials [103]. It is anticipated that these materials will soon be applied in the rapidly developing electronic information sector, including computers, networks, communications, and industrial automation.

Electronic devices in these fields extensively use light, thin, small, and highly integrated switching power supplies, which rely on high-frequency electronic technology. This necessitates that the soft magnetic cores of transformers and inductors are suitable for high-frequency applications. Bulk amorphous alloys, with their high saturation magnetic induction, high permeability, low loss, and ease of processing, can be directly cast or processed into micro-cores of various complex structures. These cores can then be made into transformers or inductors for use in various electronic or communication devices. Consequently, the application prospects and market potential for these materials are very broad.

6. Challenges and Future Perspectives

6.1. Scalability and Cost Issues in the Production of Amorphous Alloys

The production of amorphous alloys primarily relies on rapid cooling technologies, which are essential to prevent the formation of crystalline structures. This process requires complex and expensive equipment that is effective only within limited shapes and sizes, which significantly restricts the feasibility of large-scale production. Techniques such as investment casting and rapid solidification methods demand precise control over temperature and cooling rates to ensure the formation of amorphous structures. Any slight deviation in these parameters can cause crystallization, thereby compromising the material's performance and reliability.

The high cost of amorphous alloys primarily arises from the expensive, high-purity raw materials required, such as titanium, zirconium, and nickel, and the complexity of the production process. The production process itself, notably the rapid cooling technologies, demands high energy consumption and precision equipment, thereby further escalating costs. Additionally, the strict requirements for precise control and specific environmental conditions during manufacturing can cause minor issues that significantly increase the scrap rate, thus driving up overall production costs.

6.2. Optimization of Crystallization Tendency and Thermal Stability

Amorphous alloys are susceptible to crystallization under certain conditions, particularly at high temperatures or during extended use. This crystallization process can significantly deteriorate the material's properties, including strength, toughness, and corrosion resistance. Consequently, optimizing the resistance to crystallization in amorphous alloys has become a critical area of research. Adjusting the alloy's composition and processing parameters has been shown to effectively enhance its resistance to crystallization. For instance, incorporating a suitable amount of rare earth elements or employing a multi-component alloy design can inhibit the formation and growth of crystalline nuclei, thereby maintaining the desired amorphous structure.

Thermal stability in amorphous alloys refers to their capacity to retain their amorphous structure under high-temperature conditions. This property is essential for applications that involve high-temperature structural materials and electronic components. To enhance thermal stability, strategies such as optimizing alloy composition, utilizing heat treatment processes, and developing surface coatings that resist oxidation and corrosion have been employed. Furthermore, leveraging advancements in nanotechnology has led researchers to experiment with integrating nanoparticles or nanofibers into amorphous alloys, a method that promises to further boost their thermal stability.

6.3. New Alloy Design and High-Throughput Screening Techniques

a. Challenges in New Alloy Design

Complexity of multi-element systems. Amorphous alloys are typically composed of multiple elements, resulting in complex compositions that render traditional trial-and-error methods inefficient. This complexity complicates the systematic optimization of alloy properties as the introduction of each new element can unpredictably affect the alloy's structure and performance, adding uncertainty to the design process.

Balancing performance. In new alloy design, it is crucial to balance multiple performance indicators. For example, increasing strength might reduce toughness, while enhancing corrosion resistance could affect conductivity. Achieving this multi-objective optimization necessitates precise design and rigorous control, posing a significant challenge.

Limitations of theoretical models. The existing theoretical models, while useful, have limitations in guiding new alloy designs, particularly for complex multi-element systems. Accurately predicting an alloy's glass-forming ability and physical properties continues to be a challenging endeavor.

b. Advantages of High-throughput Screening Technology

Rapid screening. High-throughput screening technology facilitates the preparation and testing of numerous alloy samples with varied compositions simultaneously, significantly reducing the research cycle. This method swiftly identifies optimal composition combinations, thereby enhancing research efficiency.

Systematic research. This technology enables researchers to systematically examine the relationship between composition and performance, aiding in the creation of comprehensive databases. Such databases are crucial for uncovering patterns in alloy composition design and guiding the development of future materials.

Data-driven optimization. Integrating high-throughput screening with machine learning algorithms allows for the extraction of valuable insights from extensive experimental data, optimizing alloy design. For instance, machine learning algorithms can pinpoint potential high-performance alloy compositions from screening data, facilitating predictions and further optimizations.

7. Conclusions

7.1. Summary of Research Achievements and Technological Advances in Amorphous Alloys

Amorphous alloys, also known as metallic glasses, have garnered significant attention due to their unique properties and broad potential applications. The research and technological advancements in this field can be summarized as follows.

a. Historical milestones and production techniques. The first amorphous alloy was produced in 1960 at Caltech, which led to the discovery that rapid cooling (in the order of millions of degrees Celsius per second) is essential to prevent crystallization. Methods such as melt spinning, solid-state reactions, and mechanical alloying have been developed to produce amorphous metals. Advances in these techniques have allowed the production of bulk metallic glasses (BMGs) that can form thicker layers, improving their usability in various applications.

b. Material properties. Amorphous alloys exhibit superior mechanical properties, including high tensile strength, high elasticity, and excellent wear resistance compared with their crystalline counterparts. They also display unique electrical properties, such as high electrical conductivity and specific behaviors under temperature variations. Their magnetic properties make them suitable for applications like transformer cores and electronic article surveillance systems.

c. Applications. The special properties of amorphous alloys have been harnessed in various fields, including sports equipment, medical devices, and electronics. They are used in high-efficiency transformers, sensitive flow meters, pressure sensors, and more. Recent advances in 3D printing and other additive manufacturing techniques have enabled the creation of larger and more complex amorphous metal structures, expanding their potential uses.

7.2. Future Research Directions and Technological Challenges

Despite significant advancements, the field of amorphous alloys continues to face several challenges and opportunities for future research.

a. Enhanced production methods. Developing more efficient and scalable production techniques is crucial. Current methods require extremely rapid cooling rates, which limit the size and shape of the resulting amorphous metals. Innovations in additive manufacturing, such as selective laser melting (SLM) and laser foil printing (LFP), hold promise for overcoming these limitations, enabling the production of larger and more complex structures.

b. Material optimization. Research should focus on discovering new alloy compositions that can form glasses at slower cooling rates, making the production process more practical and cost-effective. There is a need to enhance the ductility and fatigue resistance of amorphous alloys to broaden their application range, particularly in reliability-critical fields.

c. Advanced applications. Exploring the use of amorphous alloys in biomedical applications, such as bioabsorbable implants, is a promising area. Alloys that dissolve in the body at controlled rates can revolutionize medical implants and fracture fixation devices. The development of amorphous alloys with superior properties, such as the highest recorded elastic limit, opens new possibilities in defense and aerospace for materials that can withstand extreme conditions.

d. Theoretical and computational modeling. Advanced computational models, leveraging artificial intelligence and machine learning, can accelerate the discovery of new amorphous alloy compositions and predict their properties more accurately. Theoretical studies on the electronic and atomic structures of amorphous metals will deepen the understanding of their unique behaviors and guide the development of new materials with tailored properties.

In conclusion, while the field of amorphous alloys has made significant strides, ongoing research and technological innovation are essential to fully realize their potential. Addressing production challenges, optimizing material properties, and exploring new applications will drive the future success of amorphous alloys in various industries.

Funding: This research was funded by the Open Fund Project of the State Key Laboratory of Disaster Prevention and Mitigation of Ex-plosion and Impact, College of National Defense Engineering, Army Engineering University of PLA grant number LGD-SKL-202207. And this research was funded by National Natural Science Youth Foundation of China grant number 52301353.

Conflicts of Interest: Author Yuze Zhuang was employed by the company Suzhou High Speed Management Co., Ltd. The remaining authors declare that the research was conducted in the absence of any commercial or financial relationships that could be construed as a potential conflict of interest.

References

1. Suryanarayana, C.; Inoue, A. *Bulk Metallic Glasses*; CRC press: Boca Raton, FL, USA, 2017.
2. Axinte, E. Metallic glasses from "alchemy" to pure science: Present and future of design, processing and applications of glassy metals. *Mater. Des.* **2012**, *35*, 518–556. [CrossRef]
3. Johnson, W.L. Thermodynamic and kinetic aspects of the crystal to glass transformation in metallic materials. *Prog. Mater. Sci.* **1986**, *30*, 81–134. [CrossRef]
4. Sohrabi, S.; Fu, J.; Li, L.; Zhang, Y.; Li, X.; Sun, F.; Ma, J.; Wang, W.H. Manufacturing of metallic glass components: Processes, structures and properties. *Prog. Mater. Sci.* **2024**, *144*, 101283. [CrossRef]
5. Hofmann, D.C. Bulk metallic glasses and their composites: A brief history of diverging fields. *J. Mater.* **2013**, *2013*, 517904. [CrossRef]
6. Klement, W.; Willens, R.H.; Duwez, P.O.L. Non-crystalline structure in solidified gold–silicon alloys. *Nature* **1960**, *187*, 869–870. [CrossRef]
7. Herzer, G. Modern soft magnets: Amorphous and nanocrystalline materials. *Acta Mater.* **2013**, *61*, 718–734. [CrossRef]
8. Inoue, A. Bulk amorphous alloys with soft and hard magnetic properties. *Mater. Sci. Eng. A* **1997**, *226*, 357–363. [CrossRef]
9. Kui, H.W.; Greer, A.L.; Turnbull, D. Formation of bulk metallic glass by fluxing. *Appl. Phys. Lett.* **1984**, *45*, 615–616. [CrossRef]
10. Peker, A.; Johnson, W.L. A highly processable metallic glass: Zr41. 2Ti13. 8Cu12. 5Ni10. 0Be22. 5. *Appl. Phys. Lett.* **1993**, *63*, 2342–2344. [CrossRef]
11. Zhang, B.; Zhao, D.Q.; Pan, M.X.; Wang, W.H.; Greer, A.L. Amorphous metallic plastic. *Phys. Rev. Lett.* **2005**, *94*, 205502. [CrossRef]
12. Bardt, J.A.; Bourne, G.R.; Schmitz, T.L.; Ziegert, J.C.; Gregory Sawyer, W. Micromolding three-dimensional amorphous metal structures. *J. Mater. Res.* **2007**, *22*, 339–343. [CrossRef]
13. Kumar, G.; Tang, H.X.; Schroers, J. Nanomoulding with amorphous metals. *Nature* **2009**, *457*, 868–872. [CrossRef]
14. Greer, A.L.; Ma, E. Bulk metallic glasses: At the cutting edge of metals research. *MRS Bull.* **2007**, *32*, 611–619. [CrossRef]
15. Halim, Q.; Mohamed, N.A.N.; Rejab, M.R.M.; Naim, W.N.W.A.; Ma, Q. Metallic glass properties, processing method and development perspective: A review. *Int. J. Adv. Manuf. Technol.* **2021**, *112*, 1231–1258. [CrossRef]
16. Fukushige, T.; Hata, S.; Shimokohbe, A. A MEMS conical spring actuator array. *J. Microelectromechanical Syst.* **2005**, *14*, 243–253. [CrossRef]
17. Schroers, J. Processing of bulk metallic glass. *Adv. Mater.* **2010**, *22*, 1566–1597. [CrossRef] [PubMed]
18. Peng, X.; Han, J.; Wang, Y.; Bo, Z.; Nie, A.; Li, P.; Li, Y.; Wu, H.; Liu, P.; Lu, Z.; et al. Unexpected enhanced catalytic performance via highly dense interfaces in ultra-fine amorphous-nanocrystalline biphasic structure. *Appl. Mater. Today* **2022**, *29*, 101689. [CrossRef]
19. Telford, M. The case for bulk metallic glass. *Mater. Today* **2004**, *7*, 36–43. [CrossRef]
20. Gao, K.; Zhu, X.G.; Chen, L.; Li, W.H.; Xu, X.; Pan, B.T.; Li, W.R.; Zhou, W.H.; Li, L.; Huang, W.; et al. Recent development in the application of bulk metallic glasses. *J. Mater. Sci. Technol.* **2022**, *131*, 115–121. [CrossRef]
21. Miracle, D.B. A physical model for metallic glass structures: An introduction and update. *Jom* **2012**, *64*, 846–855. [CrossRef]
22. Wang, P.F.; Jiang, H.Y.; Shi, J.A.; Liu, M.; Gu, L.; Zhou, W.; Orava, J.; Sun, Y.H.; Wang, W.H.; Bai, H.Y. Regulated color-changing metallic glasses. *J. Alloys Compd.* **2021**, *876*, 160139. [CrossRef]
23. Lee, S.; Kim, S.W.; Ghidelli, M.; An, H.S.; Jang, J.; Bassi, A.L.; Li, S.Y.; Park, J.U. Integration of transparent supercapacitors and electrodes using nanostructured metallic glass films for wirelessly rechargeable, skin heat patches. *Nano Lett.* **2020**, *20*, 4872–4881. [CrossRef] [PubMed]
24. Yan, W.; Richard, I.; Kurtuldu, G.; James, N.D.; Schiavone, G.; Squair, J.W.; Nguyen-Dang, T.; Das Gupta, T.; Qu, Y.; Cao, J.D.; et al. Structured nanoscale metallic glass fibres with extreme aspect ratios. *Nat. Nanotechnol.* **2020**, *15*, 875–882. [CrossRef]
25. Lin, X.H.; Johnson, W.L. Formation of Ti–Zr–Cu–Ni bulk metallic glasses. *J. Appl. Phys.* **1995**, *78*, 6514–6519. [CrossRef]
26. Wang, W.H.; Dong, C.; Shek, C.H. Bulk metallic glasses. *Mater. Sci. Eng. R Rep.* **2004**, *44*, 45–89. [CrossRef]
27. Wang, W.H. The elastic properties, elastic models and elastic perspectives of metallic glasses. *Prog. Mater. Sci.* **2012**, *57*, 487–656. [CrossRef]
28. Schroers, J. Bulk metallic glasses. *Phys. Today* **2013**, *66*, 32–37. [CrossRef]
29. Chen, H.S.; Miller, C.E. A rapid quenching technique for the preparation of thin uniform films of amorphous solids. *Rev. Sci. Instrum.* **1970**, *41*, 1237–1238. [CrossRef]
30. Turnbull, D. Kinetics of solidification of supercooled liquid mercury droplets. *J. Chem. Phys.* **1952**, *20*, 411–424. [CrossRef]

31. Becker, R.; Döring, W. Kinetische behandlung der keimbildung in übersättigten dämpfen. *Ann. Phys.* **1935**, *416*, 719–752. [CrossRef]
32. Dyre, J.C. Colloquium: The glass transition and elastic models of glass-forming liquids. *Rev. Mod. Phys.* **2006**, *78*, 953. [CrossRef]
33. Egami, T.; Waseda, Y. Atomic size effect on the formability of metallic glasses. *J. Non-Cryst. Solids* **1984**, *64*, 113–134. [CrossRef]
34. Suryanarayana, C. *Non-Equilibrium Processing of Materials*; Elsevier: Amsterdam, The Netherlands, 1999.
35. Zhao, R.; Jiang, H.Y.; Luo, P.; Sun, Y.T.; Li, Z.A.; Wu, W.W.; Shen, L.Q.; Liu, M.; Zhao, S.F.; Wen, P.; et al. A facile strategy to produce monatomic tantalum metallic glass. *Appl. Phys. Lett.* **2020**, *117*, 131903. [CrossRef]
36. Zhong, L.; Wang, J.; Sheng, H.; Zhang, Z.; Mao, S.X. Formation of monatomic metallic glasses through ultrafast liquid quenching. *Nature* **2014**, *512*, 177–180. [CrossRef] [PubMed]
37. Wang, Z.J.; Li, M.X.; Yu, J.H.; Ge, X.B.; Liu, Y.H.; Wang, W.H. Low-iridium-content IrNiTa metallic glass films as intrinsically active catalysts for hydrogen evolution reaction. *Adv. Mater.* **2020**, *32*, 1906384. [CrossRef]
38. Jacobson, L.A.; McKittrick, J. Rapid solidification processing. *Mater. Sci. Eng. R Rep.* **1994**, *11*, 355–408. [CrossRef]
39. Soare, V.; Burada, M.; Constantin, I.; Mitrică, D.; Bădiliţă, V.; Caragea, A.; Târcolea, M. Electrochemical deposition and microstructural characterization of AlCrFeMnNi and AlCrCuFeMnNi high entropy alloy thin films. *Appl. Surf. Sci.* **2015**, *358*, 533–539. [CrossRef]
40. Greer, A.L. Metallic glasses. *Science* **1995**, *267*, 1947–1953. [CrossRef]
41. Liebermann, H.H. The dependence of the geometry of glassy alloy ribbons on the chill block melt-spinning process parameters. *Mater. Sci. Eng.* **1980**, *43*, 203–210. [CrossRef]
42. Drehman, A.J.; Greer, A.L.; Turnbull, D. Bulk formation of a metallic glass: Pd40Ni40P20. *Appl. Phys. Lett.* **1982**, *41*, 716–717. [CrossRef]
43. Brenner, A. *Electrodeposition of Alloys: Principles and Practice*; Elsevier: Amsterdam, The Netherlands, 2013.
44. Brenner, A.; Couch, D.E.; Williams, E.K. Electrodeposition of alloys of phosphorus with nickel or cobalt. *J. Res. Nat. Bur. Stand* **1950**, *44*, 109. [CrossRef]
45. Mukhopadhyay, A.; Jangid, M.K. Li metal battery, heal thyself. *Science* **2018**, *359*, 1463. [CrossRef] [PubMed]
46. Liu, D.H.; Bai, Z.; Li, M.; Yu, A.; Luo, D.; Liu, W.; Yang, L.; Lu, J.; Amine, K.; Chen, Z. Developing high safety Li-metal anodes for future high-energy Li-metal batteries: Strategies and perspectives. *Chem. Soc. Rev.* **2020**, *49*, 5407–5445. [CrossRef] [PubMed]
47. Pedraza, D.F. Irradiation as a tool for studying solid-state amorphization phenomena. *Metall. Trans. A* **1990**, *21*, 1809–1815. [CrossRef]
48. Li, S.; Pastewka, L.; Gumbsch, P. Glass formation by severe plastic deformation of crystalline Cu | Zr nano-layers. *Acta Mater.* **2019**, *165*, 577–586. [CrossRef]
49. Saito, Y.; Tsuji, N.; Utsunomiya, H.; Sakai, T.; Hong, R.G. Ultra-fine grained bulk aluminum produced by accumulative roll-bonding (ARB) process. *Scr. Mater.* **1998**, *39*, 1221–1227. [CrossRef]
50. Suryanarayana, C. Mechanical alloying and milling. *Prog. Mater. Sci.* **2001**, *46*, 1–184. [CrossRef]
51. Jang, D.; Greer, J.R. Transition from a strong-yet-brittle to a stronger-and-ductile state by size reduction of metallic glasses. *Nat. Mater.* **2010**, *9*, 215–219. [CrossRef]
52. Greer, J.R.; De Hosson, J.T.M. Plasticity in small-sized metallic systems: Intrinsic versus extrinsic size effect. *Prog. Mater. Sci.* **2011**, *56*, 654–724. [CrossRef]
53. Bignoli, F.; Djemia, P.; Terraneo, G.; Abadias, G.; Gammer, C.; Lassnig, A.; Teixeira, C.A.; Lee, S.; Ahmadian, A.; Bassi, A.L.; et al. Novel class of crystal/glass ultrafine nanolaminates with large and tunable mechanical properties. *ACS Appl. Mater. Interfaces* **2024**, *16*, 35686–35696. [CrossRef]
54. Chu, J.P.; Jang, J.S.C.; Huang, J.C.; Chou, H.S.; Yang, Y.; Ye, J.C.; Wang, Y.C.; Lee, J.W.; Liu, F.X.; Liaw, P.K.; et al. Thin film metallic glasses: Unique properties and potential applications. *Thin Solid Film.* **2012**, *520*, 5097–5122. [CrossRef]
55. Tory, E.M.; Church, B.H.; Tam, M.K.; Ratner, M. Simulated random packing of equal spheres. *Can. J. Chem. Eng.* **1973**, *51*, 484–493. [CrossRef]
56. Kauzmann, W. The nature of the glassy state and the behavior of liquids at low temperatures. *Chem. Rev.* **1948**, *43*, 219–256. [CrossRef]
57. Miracle, D.B. A structural model for metallic glasses. *Nat. Mater.* **2004**, *3*, 697–702. [CrossRef] [PubMed]
58. Miracle, D.B.; Lord, E.A.; Ranganathan, S. Candidate atomic cluster configurations in metallic glass structures. *Mater. Trans.* **2006**, *47*, 1737–1742. [CrossRef]
59. Miracle, D.B.; Sanders, W.S.; Senkov, O.N. The influence of efficient atomic packing on the constitution of metallic glasses. *Philos. Mag.* **2003**, *83*, 2409–2428. [CrossRef]
60. Miracle, D.B. The efficient cluster packing model–An atomic structural model for metallic glasses. *Acta Mater.* **2006**, *54*, 4317–4336. [CrossRef]
61. Miracle, D.B.; Egami, T.; Flores, K.M.; Kelton, K.F. Structural aspects of metallic glasses. *Mrs Bull.* **2007**, *32*, 629–634. [CrossRef]
62. Miracle, D.B.; Louzguine-Luzgin, D.V.; Louzguina-Luzgina, L.V.; Inoue, A. An assessment of binary metallic glasses: Correlations between structure, glass forming ability and stability. *Int. Mater. Rev.* **2010**, *55*, 218–256. [CrossRef]
63. Cheng, Y.Q.; Ma, E. Atomic-level structure and structure–property relationship in metallic glasses. *Prog. Mater. Sci.* **2011**, *56*, 379–473. [CrossRef]

64. Li, M.X.; Zhao, S.F.; Lu, Z.; Hirata, A.; Wen, P.; Bai, H.Y.; Chen, M.; Schroers, J.; Liu, Y.; Wang, W.-H. High-temperature bulk metallic glasses developed by combinatorial methods. *Nature* **2019**, *569*, 99–103. [CrossRef] [PubMed]
65. Hu, L.; Bian, X.; Wang, W.; Zhang, J.; Jia, Y. Liquid fragility and characteristic of the structure corresponding to the prepeak of AlNiCe amorphous alloys. *Acta Mater.* **2004**, *52*, 4773–4781. [CrossRef]
66. Lee, G.W.; Gangopadhyay, A.K.; Hyers, R.W.; Rathz, T.J.; Rogers, J.R.; Robinson, D.S.; Goldman, A.I.; Kelton, K.F. Local structure of equilibrium and supercooled Ti-Zr-Ni liquids. *Phys. Rev. B* **2008**, *77*, 184102. [CrossRef]
67. Doye, J.P.; Miller, M.A.; Wales, D.J. Evolution of the potential energy surface with size for Lennard-Jones clusters. *J. Chem. Phys.* **1999**, *111*, 8417–8428. [CrossRef]
68. Yin, H.; Huang, Y.; Daisenberg, D.; Xue, P.; Jiang, S.; Ru, W.; Jiang, S.; Bao, Y.; Bian, X.; Tong, X.; et al. Atomic structure evolution of high entropy metallic glass microwires at cryogenic temperature. *Scr. Mater.* **2019**, *163*, 29–33. [CrossRef]
69. Fujita, T.; Konno, K.; Zhang, W.; Kumar, V.; Matsuura, M.; Inoue, A.; Sakurai, T.; Chen, M.W. Atomic-Scale Heterogeneity of a Multicomponent Bulk Metallic Glass with Excellent Glass Forming Ability. *Phys. Rev. Lett.* **2009**, *103*, 075502. [CrossRef]
70. Wochner, P.; Gutt, C.; Autenrieth, T.; Demmer, T.; Bugaev, V.; Ortiz, A.D.; Duri, A.; Zontone, F.; Grübel, G.; Dosch, H. X-ray cross correlation analysis uncovers hidden local symmetries in disordered matter. *Proc. Natl. Acad. Sci. USA* **2009**, *106*, 11511–11514. [CrossRef]
71. Billinge, S.J.; Levin, I. The problem with determining atomic structure at the nanoscale. *Science* **2007**, *316*, 561–565. [CrossRef]
72. Tian, L.; Volkert, C.A. Measuring structural heterogeneities in metallic glasses using transmission electron microscopy. *Metals* **2018**, *8*, 1085. [CrossRef]
73. Su, Z.; Decencière, E.; Nguyen, T.T.; El-Amiry, K.; De Andrade, V.; Franco, A.A.; Demortière, A. Artificial neural network approach for multiphase segmentation of battery electrode nano-CT images. *npj Comput. Mater.* **2022**, *8*, 30. [CrossRef]
74. Stoica, M.; Sarac, B.; Spieckermann, F.; Wright, J.; Gammer, C.; Han, J.; Gostin, P.F.; Eckert, J.; Löffler, J.F. X-ray Diffraction Computed Nanotomography Applied to Solve the Structure of Hierarchically Phase-Separated Metallic Glass. *ACS Nano* **2021**, *15*, 2386–2398. [CrossRef]
75. Huang, B.; Ge, T.P.; Liu, G.L.; Luan, J.H.; He, Q.F.; Yuan, Q.X.; Huang, W.X.; Zhang, K.; Bai, H.Y.; Shek, C.H.; et al. Density fluctuations with fractal order in metallic glasses detected by synchrotron X-ray nano-computed tomography. *Acta Mater.* **2018**, *155*, 69–79. [CrossRef]
76. Cao, K.; Skowron, S.T.; Biskupek, J.; Stoppiello, C.T.; Leist, C.; Besley, E.; Khlobystov, A.N.; Kaiser, U. Imaging an unsupported metal–metal bond in dirhenium molecules at the atomic scale. *Sci. Adv.* **2020**, *6*, eaay5849. [CrossRef]
77. Liu, C.; Cui, J.; Cheng, Z.; Zhang, B.; Zhang, S.; Ding, J.; Yu, R.; Ma, E. Direct Observation of Oxygen Atoms Taking Tetrahedral Interstitial Sites in Medium-Entropy Body-Centered-Cubic Solutions. *Adv. Mater.* **2023**, *35*, 2209941. [CrossRef] [PubMed]
78. Hirata, A.; Hirotsu, Y.; Ohkubo, T.; Tanaka, N.; Nieh, T.G. Local atomic structure of Pd–Ni–P bulk metallic glass examined by high-resolution electron microscopy and electron diffraction. *Intermetallics* **2006**, *14*, 903–907. [CrossRef]
79. Zhu, F.; Hirata, A.; Liu, P.; Song, S.; Tian, Y.; Han, J.; Fujita, T.; Chen, M. Correlation between local structure order and spatial heterogeneity in a metallic glass. *Phys. Rev. Lett.* **2017**, *119*, 215501. [CrossRef] [PubMed]
80. Lau, M.T.; Lan, S.; Yip, Y.L.; Kui, H.W. A metastable liquid state miscibility gap in undercooled Pd–Ni–P melts. *J. Non-Cryst. Solids* **2012**, *358*, 2667–2673. [CrossRef]
81. Hirata, A.; Kang, L.J.; Fujita, T.; Klumov, B.; Matsue, K.; Kotani, M.; Yavari, A.R.; Chen, M.W. Geometric frustration of icosahedron in metallic glasses. *Science* **2013**, *341*, 376–379. [CrossRef]
82. Hirata, A.; Guan, P.; Fujita, T.; Hirotsu, Y.; Inoue, A.; Yavari, A.R.; Sakurai, T.; Chen, M. Direct observation of local atomic order in a metallic glass. *Nat. Mater.* **2011**, *10*, 28–33. [CrossRef]
83. Miao, J.; Ercius, P.; Billinge, S.J. Atomic electron tomography: 3D structures without crystals. *Science* **2016**, *353*, aaf2157. [CrossRef]
84. Yang, Y.; Zhou, J.; Zhu, F.; Yuan, Y.; Chang, D.J.; Kim, D.S.; Pham, M.; Rana, A.; Tian, X.; Yao, Y.; et al. Determining the three-dimensional atomic structure of an amorphous solid. *Nature* **2021**, *592*, 60–64. [CrossRef]
85. Yuan, Y.; Kim, D.S.; Zhou, J.; Chang, D.J.; Zhu, F.; Nagaoka, Y.; Yang, Y.; Pham, M.; Osher, S.J.; Chen, O.; et al. Three-dimensional atomic packing in amorphous solids with liquid-like structure. *Nat. Mater.* **2022**, *21*, 95–102. [CrossRef] [PubMed]
86. Inoue, A.; Shen, B.L.; Koshiba, H.; Kato, H.; Yavari, A.R. Ultra-high strength above 5000 MPa and soft magnetic properties of Co–Fe–Ta–B bulk glassy alloys. *Acta Mater.* **2004**, *52*, 1631–1637. [CrossRef]
87. Demetriou, M.D.; Launey, M.E.; Garrett, G.; Schramm, J.P.; Hofmann, D.C.; Johnson, W.L.; Ritchie, R.O. A damage-tolerant glass. *Nat. Mater.* **2011**, *10*, 123–128. [CrossRef]
88. Zhang, J.; Liu, Y.; Yang, H.; Ren, Y.; Cui, L.; Jiang, D.; Wu, Z.; Ma, Z.; Guo, F.; Bakhtiari, S.; et al. Achieving 5.9% elastic strain in kilograms of metallic glasses: Nanoscopic strain engineering goes macro. *Mater. Today* **2020**, *37*, 18–26. [CrossRef]
89. Sun, Y.; Zhang, H.F.; Fu, H.M.; Wang, A.M.; Hu, Z.Q. Mg–Cu–Ag–Er bulk metallic glasses with high glass forming ability and compressive strength. *Mater. Sci. Eng. A* **2009**, *502*, 148–152. [CrossRef]
90. Gong, P.; Wang, X.; Shao, Y.; Chen, N.; Liu, X.; Yao, K.F. A Ti–Zr–Be–Fe–Cu bulk metallic glass with superior glass-forming ability and high specific strength. *Intermetallics* **2013**, *43*, 177–181. [CrossRef]
91. Tan, J.; Pan, F.S.; Zhang, Y.; Sun, B.A.; He, J.; Zheng, N.; Stoica, M.; Kühn, U.; Eckert, J. Formation of Zr–Co–Al bulk metallic glasses with high strength and large plasticity. *Intermetallics* **2012**, *31*, 282–286. [CrossRef]
92. Liu, Y.H.; Wang, G.; Pan, M.X.; Yu, P.; Zhao, D.Q.; Wang, W.H. Deformation behaviors and mechanism of Ni–Co–Nb–Ta bulk metallic glasses with high strength and plasticity. *J. Mater. Res.* **2007**, *22*, 869–875. [CrossRef]

93. Wang, J.; Cao, W.; Wang, L.; Zhu, S.; Guan, S.; Huang, L.; Li, R.; Zhang, T. Fe–Al–P–C–B bulk metallic glass with good mechanical and soft magnetic properties. *J. Alloys Compd.* **2015**, *637*, 5–9. [CrossRef]
94. Inoue, A.; Shen, B.L.; Chang, C.T. Fe-and Co-based bulk glassy alloys with ultrahigh strength of over 4000 MPa. *Intermetallics* **2006**, *14*, 936–944. [CrossRef]
95. Xi, X.K.; Zhao, D.Q.; Pan, M.X.; Wang, W.H.; Wu, Y.; Lewandowski, J.J. Fracture of brittle metallic glasses: Brittleness or plasticity *Phys. Rev. Lett.* **2005**, *94*, 125510. [CrossRef] [PubMed]
96. Gilbert, C.J.; Schroeder, V.; Ritchie, R.O. Mechanisms for fracture and fatigue-crack propagation in a bulk metallic glass. *Metall. Mater. Trans. A* **1999**, *30*, 1739–1753. [CrossRef]
97. Inoue, A. High strength bulk amorphous alloys with low critical cooling rates (overview). Materials Transactions. *JIM* **1995**, *36* 866–875.
98. Révész, Á.; Kis-Tóth, Á.; Varga, L.K.; Schafler, E.; Bakonyi, I.; Spassov, T. Hydrogen storage of melt-spun amorphous Mg65Ni20Cu5Y10 alloy deformed by high-pressure torsion. *Int. J. Hydrog. Energy* **2012**, *37*, 5769–5776. [CrossRef]
99. Lang, X.Y.; Guo, H.; Chen, L.Y.; Kudo, A.; Yu, J.S.; Zhang, W.; Inoue, A.; Chen, M.W. Novel nanoporous Au− Pd alloy with high catalytic activity and excellent electrochemical stability. *J. Phys. Chem. C* **2010**, *114*, 2600–2603. [CrossRef]
100. Cao, J.D.; Kirkland, N.T.; Laws, K.J.; Birbilis, N.; Ferry, M. Ca–Mg–Zn bulk metallic glasses as bioresorbable metals. *Acta Biomater* **2012**, *8*, 2375–2383. [CrossRef] [PubMed]
101. Guo, S.F.; Zhang, H.J.; Liu, Z.; Chen, W.; Xie, S.F. Corrosion resistances of amorphous and crystalline Zr-based alloys in simulated seawater. *Electrochem. Commun.* **2012**, *24*, 39–42. [CrossRef]
102. Grimberg, A.; Baur, H.; Bochsler, P.; Bühler, F.; Burnett, D.S.; Hays, C.C.; Heber, V.S.; Jurewicz, A.J.G.; Wieler, R. Solar wind neon from Genesis: Implications for the lunar noble gas record. *Science* **2006**, *314*, 1133–1135. [CrossRef]
103. Inoue, A. *New Functional Materials, Fundamentals of Metallic Glasses and Their Applications to Industry*; Technos Co.: Toyokawa-shi Japan, 2009; ISBN 978-4-914728-58-5.

inorganics

MDPI

Article

Nickel Ions Activated PbO–GeO_2 Glasses for the Application of Electrolytes and Photonic Devices

L. Vijayalakshmi [1,*], Shaik Meera Saheb [2], R. Vijay [3], Kishor Palle [4], P. Ramesh Babu [5], Seong-Jin Kwon [1,*] and G. Naga Raju [6]

1 Department of Automotive Engineering, Yeungnam University, Gyeongsan-si 38541, Republic of Korea
2 Department of Physics, Sai Spurthi Institute of Technology, B. Gangaram, Sathupally Mandal 507303, India
3 Department of Physics, Usha Rama College of Engineering & Technology, Telaprolu 521109, India
4 Department of H&S (Chemistry), CVR College of Engineering, Hyderabad 501510, India
5 Department of Physics, Gokaraju Rangaraju Institute of Engineering & Technology, Hyderabad 500090, India
6 Department of Physics, Acharya Nagarjuna University, Nagarjuna Nagar, Guntur 522510, India
* Correspondence: lvl.phy@gmail.com or drlvlphy@ynu.ac.kr (L.V.); sjkwon@yu.ac.kr (S.-J.K.)

Abstract: In this study, PbO–GeO_2 glasses were melt-quenched at different nickel oxide concentrations. XRD and DSC techniques were characterized whether the samples are glass or crystalline materials. IR, Raman, and optical absorption techniques are used to obtain structural details. The IR spectra have revealed that the glass network contained conventional structural units GeO_4 and GeO_6. The Ni^{2+} ion octahedral transition exhibited luminescence spectra in the region of 1200–1500 nm; it is due to $^3T_2$ (3F) → $^3A_2$(3F) transition. The glasses containing the highest concentration of NiO have been found to have high values of luminescence efficiency and the cross-section. The dielectric characteristics, such as the dielectric constant, loss, and a.c. conductivity (σ_{ac}), were analyzed across extensive frequency and temperature ranges, with a specific emphasis on the nickel oxide concentration. Analyzing optical absorption and dielectric properties of the samples, it has been found that nickel ions' majority occur in tetrahedral sites. It is proved that the dielectric constant and loss values are highest for the sample N_{10} and ac conductivity due to dipoles being lowest for the sample N_{10}. It is revealed that the glasses are highly conducting due to the modifying action of Ni^{2+} ions so these glasses are suitable for solid electrolyte uses besides their optical applications in NLO devices.

Keywords: lead germanium nickel oxide glasses; spectroscopic properties; dielectric properties

1. Introduction

Glass plays a crucial role in various applications and technologies, making it an essential material in our daily lives. So, the glass industry was founded with the aim of creating a wide range of compounds for everyday use. In all these materials, the most important heavy metal oxide PbO in general behaves as a glass modifier. Upon entering into a glass network, the structure of the material changes. The low crystallization value and high moisture resistance are exposed by this information [1]. It has both a covalent and an ionic nature, which allows it to function as both a glass maker and a modifier [2]. The refractive index and melting point of materials are enhanced when lead oxide is introduced to the glass former. These types of glasses are utilized in situations where shielding, dispersion, and high refractive index are necessary [3]. The glassy nature is enhanced by [PbO_4] structural units whereas the network modifying action is due to [PbO_6] structural units of PbO. However, germanium oxide is commonly encountered in its vitreous state. It is considered a classic glass former and is known for its exceptional optical characteristics, including a refractive index (n~2.17). Moreover, it serves as an efficient transmitter in the near-infrared (NIR) region [4]. Germanium oxide glasses are used for making optical fibers and designing laser devices and in the field of nonlinear optics [5]. Since GeO_2 glasses have a high ionic conductivity, solid electrolytes are usually composed of them [6]. A structural analysis of germanium oxide indicates that it is composed of structural units made up of GeO_4 and GeO_6 [7–9]. Depending on the type of bond between lead and oxygen atoms, the modifier PbO will act either as a network former or a modifier. Because of Pb ion high polarizability, a strong covalent Pb–O bond is formed between Pb^{2+} and O^{2-} ions [10]. Lead germanate glasses are promising candidates for optical amplifiers, and for numerous types of optical devices and high-speed optical switches containing nonlinearity [11].

The transition metal ions continue to be of interest for possible laser applications. Numerous transition metal activators have shown promise, particularly those that have led to successful lasers utilizing cation matrices. The existence and level of cations in glass have a notable influence on the key physical, structural, and chemical characteristics of these materials, encompassing electrical, magnetic, and optical properties [12,13].

Nickel containing glasses gained special attraction due to their interesting optical and dielectric properties; nickel exists in highly suitable Ni^{2+} ions, which mostly occupies octahedral sites in glasses with high crystal field energy [14]. There is no accurate evidence of change in its oxidation state during glass formation and annealing [15]. Visible and near-infrared region Ni^{2+} ions can show many absorption bands. The Ni^{2+} ions in octahedral configuration in a glass matrix can exhibit lasting action at about a 1.56 μm wavelength at room temperature. This wavelength is very important in telecommunications [16]. The zero-dispersion wavelength for GeO_2 is 1.7 μm (higher than that of borate, phosphate, or silicate glasses). The emission of laser beams in the near-infrared region is because of the existence of Ni^{2+} ions in the PbO–GeO_2 glass network [17]. In addition to being used in nonlinear optical devices, these glasses were also used in the preparation of broad band optical amplifiers, power limiters, optical amplifiers, ultrafast optical switches, and lasers [18,19]. The coordination of Ge would continuously change from GeO_4 to GeO_6 with the addition of alkali oxides. Recently, the structural properties of germanate-based glasses have been examined in Na_2O-GeO_2-TeO_2 [20], MnO-GeO_2-PbO_2 [21], TeO_2-GeO_2-PbO [22], Na_2CO_3-CaO-GeO_2 [23], and Ga_2O_3-GeO_2-BaO [24]. Among the reported germanate-based glasses, the lead germanate matrix has received the most attention due to its excellent properties, i.e., a relatively large glass-forming region, high transmittance in a wide wavelength region, superior chemical durability, and thermal stability [25,26].

Very little work was reported about nickel oxide-doped lead germanate glasses; moreover, work on the dielectric properties of dense glass systems like PbO–GeO_2 glasses is very strange. So, the present work is aimed to report spectral features and dielectric studies of a nickel oxide-doped PbO–GeO_2 glass system by means of spectroscopic techniques and electrical measurements, thereby comprehending the environment of nickel ions and explaining suitability of these glasses for specific uses as electrolytes.

2. Results

From the measured density values of PbO–GeO_2 glass samples, various other physical parameters such as polaron radius r_p and nickel ion concentration N_i are calculated and presented in Table 1. As the concentration of nickel oxide increases, the density of the samples is also increased. The structural variations in the glass network are investigated by physical properties of the samples.

Table 1. Summary of the data physical parameters of the PbO–GeO_2: NiO glasses.

Glass Sample	Density d (g/cm^3)	Molar Volume V_m (cm^3)	Nickel Ion Conc. N_i ($\times 10^{21}$/cm^3)	Inter-Ionic Distance of Nickel Ions R_i (Å)	Polaron Radius R_p (Å)
N_0	5.56	0.520	--	--	--
N_2	6.035	0.886	4.78	5.93	2.39
N_4	6.054	0.586	9.60	4.70	1.89
N_6	6.072	0.953	14.44	4.10	1.654
N_8	6.08	1.008	19.29	3.72	1.50
N_{10}	6.15	0.908	24.41	3.44	1.38

Figure 1 illustrates the amorphous structure of the samples, as demonstrated by the absence of Bragg peaks in X-ray diffraction patterns.

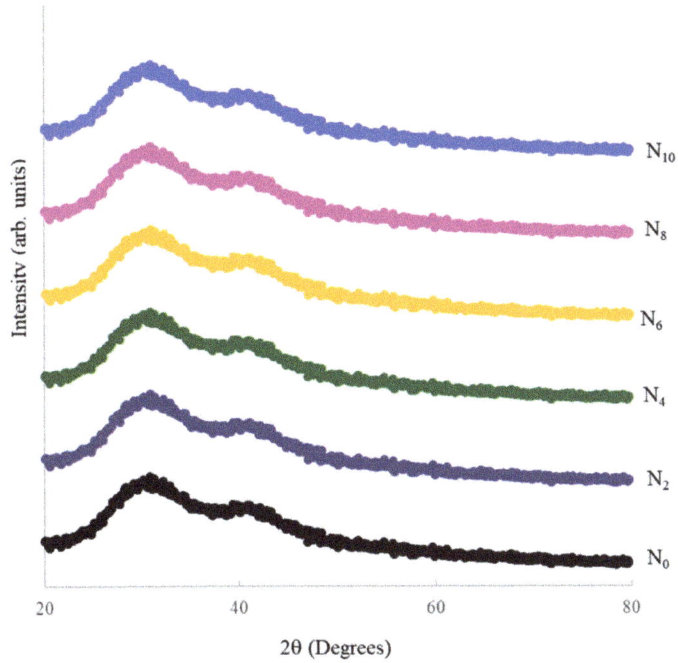

Figure 1. XRD patterns of PbO–GeO_2: NiO glasses.

Figure 2 displays the differential scanning calorimetric traces of the PbO–GeO_2 glass system doped with nickel oxide, along with the glass-forming ability parameter (K_{gl}) and the variation of glass transition temperature (T_g) as indicated in the inset. The pattern demonstrates fluctuations in T_g in the temperature range of 490 to 510 °C with an endothermic effect. Additionally, an exothermic effect is observed due to the crystallization temperature T_c, which falls in the range of 615 to 630 °C. The DSC data in Table 2 illustrate the changes in (T_g), ($T_c - T_g$), ($T_m - T_c$), and (K_{gl}) [27] of PbO–GeO_2 glasses as they are doped with varying nickel oxide concentrations. It has been observed that the glass-forming ability parameter K_{gl} decreases as the concentration of nickel oxide increases. This trend may be attributed to the alterations in the glass composition and structural properties resulting from the addition of nickel oxide.

Table 2. DSC data of PbO-GeO_2 glasses doped with different concentrations of NiO.

Glass	T_g (°C)	T_c (°C)	T_m (°C)	$T_c - T_g$ (°C)	K_{gl} = $(T_c - T_g)/(T_m - T_c)$
N_0	497	627	762	130	0.963
N_2	499	625	763	126	0.913
N_4	503	623	766	120	0.834
N_6	505	622	769	117	0.796
N_8	506	619	771	113	0.743
N_{10}	504	615	772	111	0.707

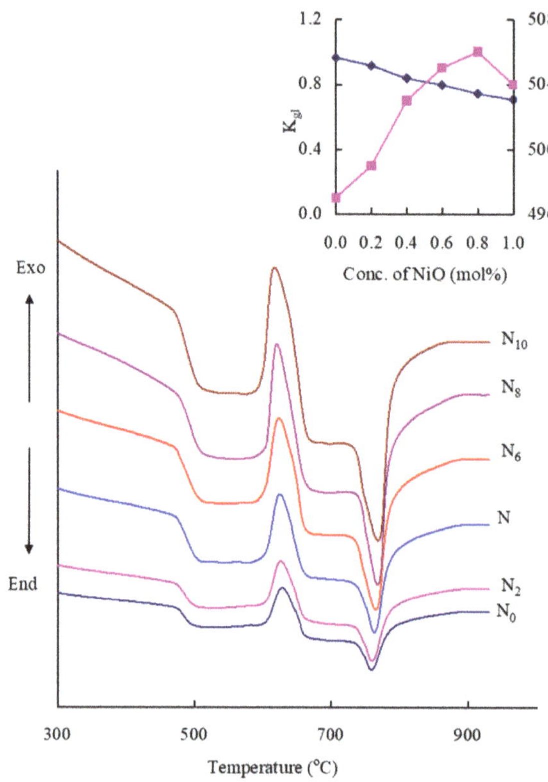

Figure 2. DSC patterns of PbO–GeO_2: NiO glasses. The inset shows the variation of T_g and K_{gl} with the concentration of NiO.

As shown in Figure 3, titled glass samples were investigated in the 350–1550 nm wavelength range at room temperature. Between 700 nm and 820 nm, a prominent band was observed in these spectra. The width and height are increased with increasing the content of NiO in these spectra. Regarding influences of Ni^{2+} ions, a number of absorption bands are observed. In the visible and NIR regions, 1.0 mol% glass (N_{10}) shows six evidently resolved absorption bands at 1265 nm (O_{h1}), 1029 nm (T_{d1}), 781 nm (O_{h2}), 623 nm (T_{d2}), 495 nm (O_{h3}), and 421 nm (O_{h4}) [28,29]. When the concentration of nickel increases, the intensity of octahedral bands is increased while tetrahedral bands are observed to decrease.

We calculated optical band gaps (E_o) of samples from cutoff wavelengths by using the relation

$$(\alpha\hbar\omega)^2 = C(\hbar\omega - Eo) \quad (1)$$

where α is the absorption coefficient (cm^{-1}), $\hbar\omega$ is photon energy of incident radiation (eV), E_o is the optical band gap energy (eV), and C is the band tailing parameter.

Figure 4 indicates that the Tauc plots and inset give optical band gap variation with concentration. Using the linear portion as a reference, E_o values are determined as shown in Table 3; the maximum E_o value is observed for the N_2 glass sample.

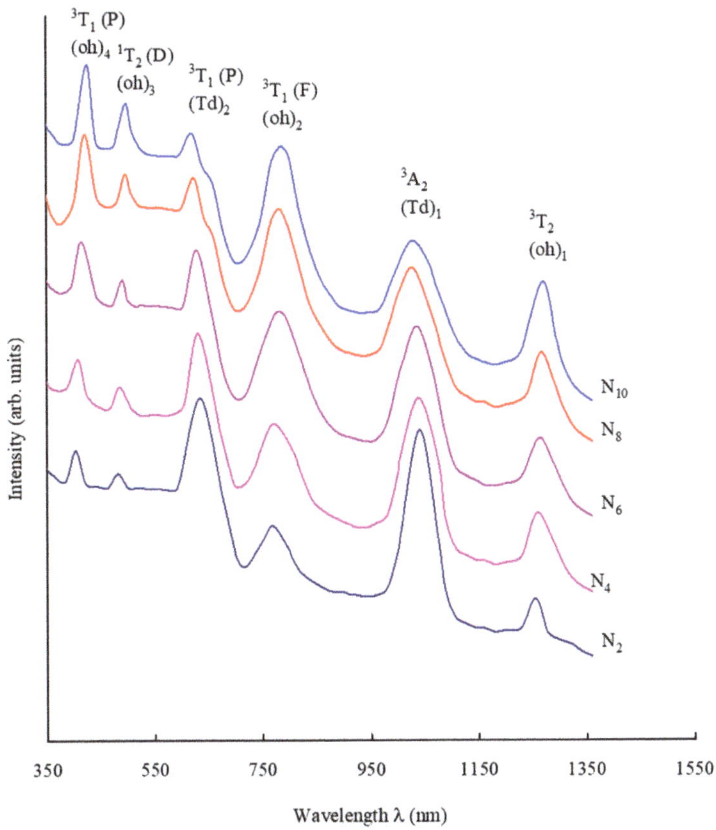

Figure 3. Optical absorption spectra of PbO-GeO$_2$ glasses doped with nickel oxide.

Table 3. Summary of the data related to optical absorption spectra of the PbO–GeO$_2$: NiO glasses.

Band Position (nm)	**N$_2$**	**N$_4$**	**N$_6$**	**N$_8$**	**N$_{10}$**
Ni^{2+} band positions, octahedral transitions (nm) $^3A_2(F) \rightarrow {}^3T_2(F)$	1254	1258	1262	1265	1269
$^3A_2(F) \rightarrow {}^3T_1(F)$	769	771	774	781	784
$^3T_2(F) \rightarrow {}^1T_2(D)$	482	484	490	495	497
$^3A_2(F) \rightarrow {}^3T_1(P)$	403	408	415	421	425
Ni^{2+} band positions, tetrahedral transitions (nm) $^3A_2(F) \rightarrow {}^3A_2(F)$	1040	1038	1034	1029	1024
$^3A_2(F) \rightarrow {}^3T_1(F)$	635	631	628	623	620
Optical band gap E_o (eV)	2.58	2.51	2.36	2.32	2.27

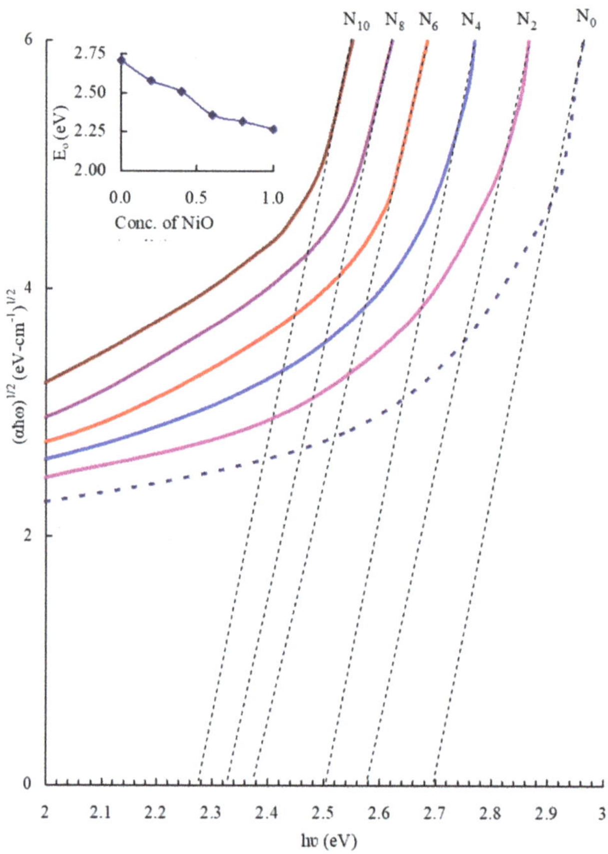

Figure 4. Tauc plots of PbO–GeO_2: NiO glasses. The inset represents the variation of the optical band gap with the concentration of NiO.

The FTIR spectra have been depicted in Figure 5. These spectra exhibit two crucial bands that correspond to the tetrahedral and octahedral vibrations of GeO_2 at 1075 and 780 cm^{-1}, respectively. Because of GeO_6 and GeO_4 units, asymmetrical stretching vibrations regarding these bands are located [18]. One more band is observed due to bonding between two tetrahedral groups at 642 cm^{-1} [19,30]. The PbO_4 tetrahedral units are located at about 448 cm^{-1} [18]. Intensity of GeO_6 and Ge–O–Ge of GeO_2 bands decreased and shifted gradually towards higher frequencies; similarly, GeO_4 and PbO_4 bands shifted towards lower frequencies with increased intensity with the introduction of nickel oxide into the glass network. The FTIR bands are shown in Table 4.

The Raman spectra of PbO–GeO_2: NiO are shown in Figure 6. These spectra contained symmetric Ge–O–Ge bands due to Q^4 units of GeO_4 in between the 420 and 450 cm^{-1} region [31], a vibrational band of Q^3 units of GeO_4 in the region 520–540 cm^{-1}, a vibrational band of GeO_6 blocks at 620 cm^{-1} [32], and two weak bands due to Q^2 and Q^1 blocks of GeO_4 at 900 and 960 cm^{-1}, respectively. PbO_4 exhibited blocks at 280 cm^{-1}. A high intense vibrational band of localized Ge–O^- stretching of the meta-germanate blocks at around 750 cm^{-1} and no bands are exhibited by NiO units. The information regarding the band positions of Raman spectra can be found in Table 5.

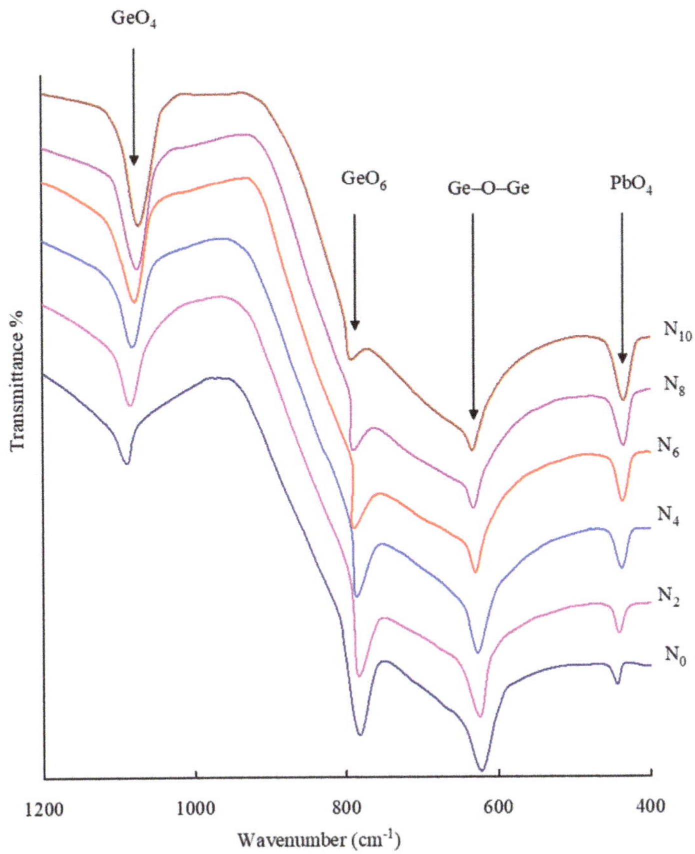

Figure 5. FTIR spectra of PbO–GeO$_2$ glasses doped with nickel oxide.

Table 4. Summary of the band positions (cm^{-1}) of various structural units in FT-IR spectra of PbO–GeO$_2$: NiO glasses.

Glass	GeO$_4$ Units	GeO$_6$ Units	Ge–O–Ge Bending Vibrations	PbO$_4$ Units
N$_0$	1088	783	622	445
N$_2$	1085	785	625	442
N$_4$	1080	787	627	439
N$_6$	1077	791	631	437
N$_8$	1075	792	633	436
N$_{10}$	1072	794	634	435

Photoluminescence spectra of PbO–GeO$_2$: NiO glasses are shown in Figure 7. Photoluminescence spectra of nickel oxide were recorded at 300 k with λ_{exc} = 800 nm. In the wavelength region 1200–1500 nm, the spectra exhibited a wider emission for all the samples due to the $^3T_2$ (3F) $\rightarrow$ $^3A_2$(3F) transition of Ni^{2+} ions. The intensities and line widths of the band increase with nickel oxide concentration as shown in Table 6, while the peaks shift toward higher wavelengths.

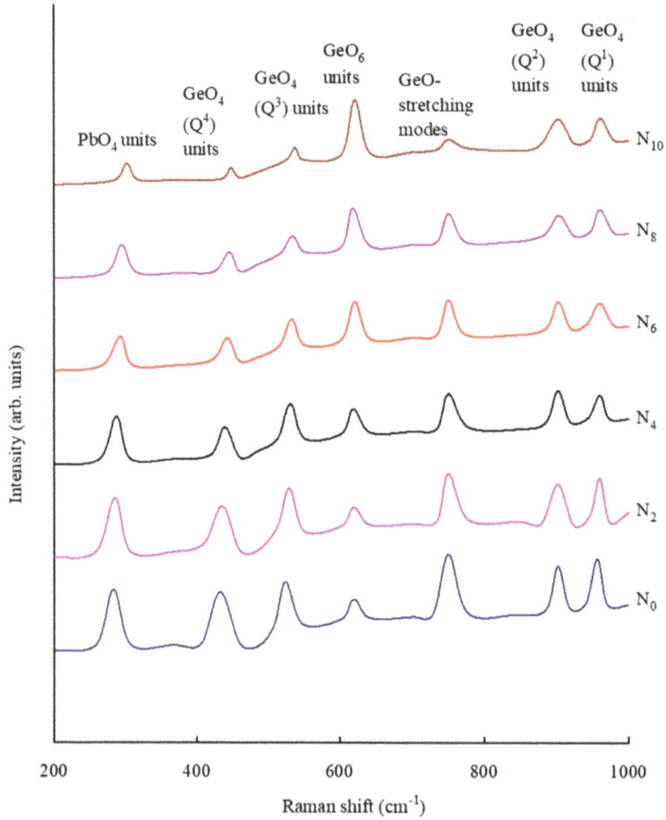

Figure 6. Raman spectra of PbO–GeO_2 glasses doped with nickel oxide.

Table 5. Summary of the data on band positions (cm^{-1}) of Raman Spectra of PbO–GeO_2: NiO glasses.

Glass	PbO_4	GeO_4 (Q^4) Units	Geo$_4$ (Q^3) Units	GeO_6 Units	GeO^- Units	GeO_4 (Q^2) Units	GeO_4 (Q^1) Units
N_0	282	433	524	621	750	901	959
N_2	284	436	529	620	750	901	959
N_4	286	440	530	619	750	901	959
N_6	292	443	532	618	750	901	959
N_8	294	446	534	618	750	901	959
N_{10}	301	448	537	618	750	901	959

Table 6. Summary of data on photoluminescence of PbO–GeO_2: NiO glasses.

Sample	Emission Peak Position (nm)	Refractive Index	σ_p^E (10^{33}, cm^2)
N_2	1303	1.542	0.932
N_4	1306	1.545	0.934
N_6	1309	1.549	0.939
N_8	1313	1.552	0.943
N_{10}	1316	1.556	0.945

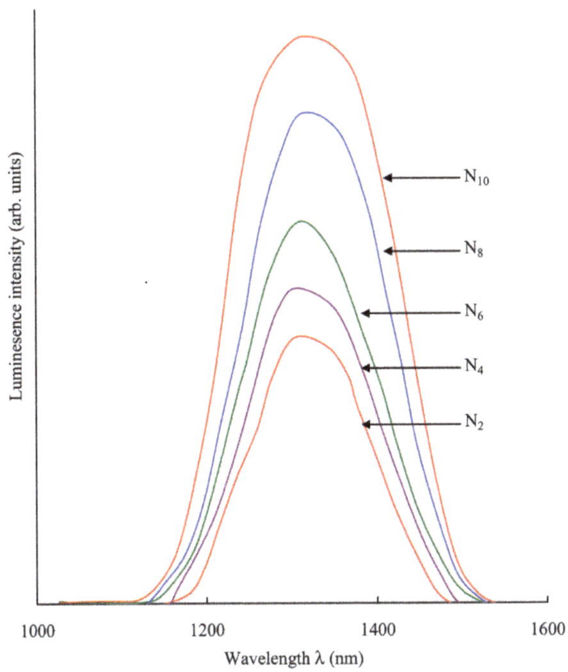

Figure 7. Photoluminescence of PbO–GeO_2 glasses doped with nickel oxide recorded at room temperature (λ_{exc} = 800 nm).

The variation of ε' verses temperature graphs of the glasses is shown in Figure 8. At higher temperature and lower frequencies, slight increases in the dielectric constant were observed and it increases abruptly with temperature. It should be high for glass N_{10}. Figure 9 shows the temperature versus tan δ graphs of glass samples at different frequencies. The curves have shown a maximum value of tan δ for the glass N_{10}. The σ_{ac} values are obtained by changing temperature using the following equation:

$$\sigma_{ac} = \omega\varepsilon^1\varepsilon_0 \tan\delta \quad (2)$$

where σ_{ac} is ac electrical conductivity (S/m), ω is angular frequency (s^{-1}), and ε^1 is the dielectric constant.

Figure 10 represents the plots of ac conductivity, σ_{ac}, with $1/T$ and is drawn at 100 kHz for all the glasses. The inset of Figure 10 explains the variation of activation energy with changes in electrical conductivity σ_{ac}. N_{10} has shown the lowest value of activation energy as presented in Table 7.

Table 7. Data on dielectric parameters of PbO-GeO_2: NiO glasses.

Glass	A.E for Conduction (eV)	σ_{ac} (10^{-7}) (Ω cm)$^{-1}$
N_0	0.43	0.56
N_2	0.39	0.77
N_4	0.36	1.08
N_6	0.33	1.59
N_8	0.29	2.54
N_{10}	0.25	4.22

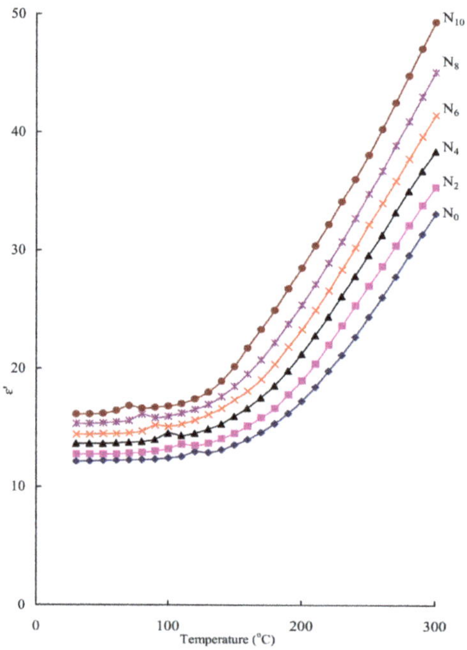

Figure 8. The dispersion of the dielectric constant, ε', vs. temperature at 100 kHz for the glasses PbO-GeO_2 doped with different concentrations of nickel oxide.

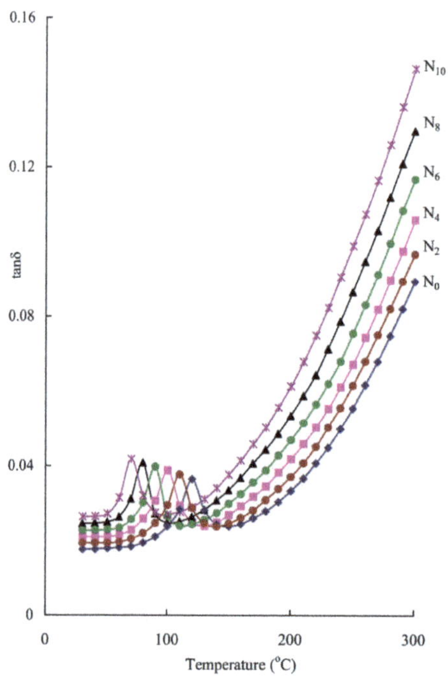

Figure 9. The variation of dielectric loss with temperature at 10 kHz frequency for the glasses PbO-GeO_2 doped with different concentrations of NiO.

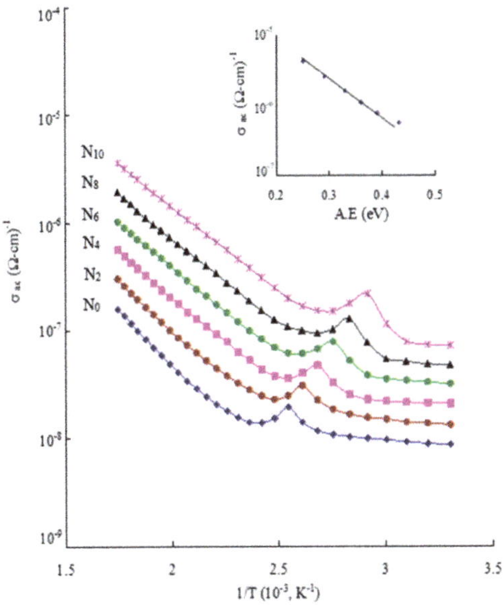

Figure 10. The variation of ac conductivity with 1/T at 100 kHz of PbO–GeO_2 glasses doped with nickel oxide. The inset represents the variation of ac conductivity regarding the activation energy.

It is verified that the σ_{ac} increased as activation energy (A.E) decreased, following the Arrhenius equation.

$$\sigma_{ac} = \sigma_0 e - {}^{E_a}/_{KT} \tag{3}$$

where σ_{ac} is ac electrical conductivity (S/m) and σ_0 is conductivity at zero Kelvin.

E_a is activation energy (eV), T is temperature, and K is the Boltzmann constant.

The variation of the $4\pi N\mu^2/27$ K value with nickel oxide concentration is shown in Figure 11.

$$\frac{\varepsilon_s - \varepsilon_\infty}{(\varepsilon_s + 2)(\varepsilon_\infty + 2)} T = \frac{4\Pi N\mu^2}{27K} \tag{4}$$

where ε_s is the low-frequency dielectric constant, ε_∞ is the high-frequency dielectric constant, $N\mu^2$ is the strength of dipoles (C^2m^{-1}), and K is the Boltzmann constant.

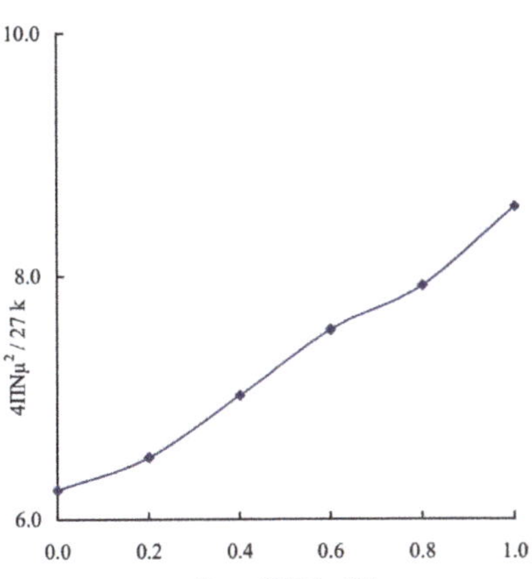

Figure 11. Dependence of $4\pi N\mu^2/27$ K on concentration of nickel oxide.

$N\mu^2$ is the strength of dipoles. For various concentrations of nickel oxide, the values of ε_s and ε_∞ subsisting in the above equation and $4\pi N\mu^2/27$ K are calculated at 393 K. The curve has shown a rising trend with nickel oxide concentration.

3. Discussion

In the PbO–GeO_2 glass system, the presence of GeO_2 in conjunction with GeO_4 tetrahedra within the network of the glass can lead to the conversion of a few tetrahedral units into GeO_6. There is typically one GeO_6 octahedron separated by one GeO_4 tetrahedron that shares one oxygen corner. Studies on GeO_2 glasses have revealed interactions between the GeO_5 structural units and germanium ions, where four bonds are of shorter length (ranging from 1.76 to 1.85 Å) and one bond has a longer interatomic distance of 1.92 Å.

The stretching of GeO_4 tetrahedra of a Ge–O bond is induced by the modifier PbO. As a result, the presence of Pb ions leads to the formation of three-membered rings in the vicinity of GeO_4 tetrahedra. The results revealed that Ge ions also participated in the glass structure with germanium dioxide network units. In addition, the infrared spectra have shown that there are bonds between GeO_4 and PbO_4, PbO_2, PbO_5, and PbO_6 structural units [33]. Lead oxide is a glass modifier; it forms Ge–O–Ge linkages by entering into the glass structure along with NBOs and dangling bonds. When lead ions form four covalent bonds with oxygen atoms, they can form structural units called [PbO_4] and [PbO_2]. [PbO_5] and [PbO_6] are produced by the lead ions and form Ge–O–Pb linkages whereas the Ni^{2+} acts as a network modifier. Transition metal ions, including Ni^{2+}, are known to influence the structural, optical, and electrical properties of glasses significantly. Ni^{2+} can act as a network modifier by occupying interstitial sites and creating non-bridging oxygens (NBOs), thus altering the glass matrix's connectivity and properties. In our study, Ni^{2+} ions are introduced into the PbO–GeO_2 glass network, primarily occupying octahedral sites. The observed spectroscopic features and the dielectric properties of the glasses support this conclusion. Our findings are consistent with previous research that highlighted similar roles for transition metal ions in glass systems [34].

Usually, the density of glass samples depends on glass structure, size of grains, coordination number of glass-forming ions, size of interstitial spaces, etc. The density of samples is increased with the content of NiO, indicating its higher compactness for the N_{10} sample.

The absence of sharp Bragg peaks indicates that the samples are in short-ranged structured materials (amorphous in nature). The glass-forming ability ($T_c - T_g$) decreased as the content of nickel oxide increased. So, it is proved that the lowest nickel oxide content is preferred for suitable glasses. It is inferred that glass samples with lower values will have less thermal stability. The decrease in interatomic distance, cross-linking, and closely packed structure are the reasons for such a behavior. Further, it is justified that the Ni^{2+} ions mostly acted as a modifier in (O_h) configuration as the NiO content is increased.

According to Tanabe–Sugano notation, the transitions of the d^8 ion of nickel are mentioned below: $^3A_2(F) \rightarrow {}^3T_2$ (F) (Oh_1), $^3T_1$ (F) (Oh_2), $^1T_2$ (D) (Oh_3). The $^3A_2(F) \rightarrow {}^1T_2$ (D) corresponds to a spin forbidden band at 570 nm while the band at 530 nm is due to $^3A_2$ (F) $\rightarrow {}^3T_1$ (P) tetrahedral transition [35]. The increasing intensity in the octahedral bands with an increase in the mol% of NiO proved the presence of octahedrally bonded Ni^{2+} ions in a large number rather than the tetrahedrally bonded Ni^{2+} ions.

The optical band gap decreases as nickel oxide concentration increases. With an increase in doping levels, Ni^{2+} ion donor centers in a large number are produced. So, the excited states of local electrons superpose with the unfilled 3d states of nearby Ni^{2+} sites. This causes the nickel oxide band to extend into the main band gap to a greater extent. From N_2 to N_{10}, the band gap has fallen due to the shift of cutoff wavelengths to a higher wavelength side.

The GeO_6 band intensity is increased with an increasing concentration of nickel oxide. This indicates that nickel ions are located in the glass structure in an octahedral fashion. In addition to the wavelength shift of the octahedral bands, nickel oxide doping also increases the wavelength shift. This shift in the band position indicated higher bond length of the Ni–O bond.

The IR spectra indicate the presence of symmetrical and anti-symmetrical bands corresponding to germanate structural units. The nickel oxide concentration is inversely proportional to the intensity of symmetric (GeO_6) units. Due to Ge–O–Ge symmetric vibrations, the intensity of bands decreases and anti-symmetrical bands due to (GeO_4) have increased as shown in Table 4. These results suggested decreasing entropy in the glass network with a hike in nickel oxide content.

It has been observed that Ge-O–Ge bonds of Q^2 and Q^1 units of GeO_4 show weak asymmetric stretching at about 900 and 960 cm^{-1} in Raman spectra. It was observed that the meta-germanate Ge–O^- unit stretching modes showed a strong vibrational band at around 750 cm^{-1}.

In the region 430 and 520 cm^{-1}, GeO_4 exhibits symmetric Ge–O–Ge bands due to the Q^4 unit and a vibrational band of the Q^3 unit. Additionally, a vibrational band of GeO_6 and PbO_4 blocks are observed at 620 and 280 cm^{-1}. No bands are exhibited by nickel oxide. The intensity of asymmetric stretching vibrations of GeO_6 units is increased at the expense of symmetric stretching vibrations regarding GeO_4 units in the Raman spectra. The concentration of nickel oxide increases and evidently suggested disorder in the glass network decreases.

In general, octahedral nickel ions produce photoluminescence, and tetrahedral nickel ions have no photoluminescence. In the green, red, and near-infrared (NIR) wavelength ranges, the octahedral Ni^{2+} ions produce luminescence bands [36]. For our samples, λ_{exc} = 800 nm and the band ranges from 1200 to 1500 nm and the meta-center shifts towards a higher wavelength with an increase in nickel oxide content. The Stokes shift increases between higher and lower bands. The value of the emission cross-section σ_p^E has been computed and tabulated. Increasing the content of nickel oxide, the value of σ_p^E is found to increase gradually from N_2 to N_{10}, proving the increase in luminescence efficiency.

Electronic, ionic, dipolar, and interfacial polarizations are the different types of polarizations. These polarizations can influence the relative permittivity ε' and interfacial polarization is a response to the structure of glasses. The value of interfacial polarization depends on the dielectric parameters (ε', tan δ). These are found to slightly increase and σ_{ac} decreases with temperature for any frequency. The activation energy for conduction decreased with the increase in nickel oxide content (Table 7), which proved the increase in the value of interfacial polarization.

The progressive rise in modifying ion concentration generates wrong bonds in the glass network. These defects cause rather easy movement of charges, thereby enhancing interfacial polarization and hence increasing the values of dielectric parameters.

Figure 10 depicts the variation of ac conductivity with the content of nickel oxide; the curves produced a maximum for pure glass and minimum for x = 1.0 mol%. The trend of the curves justified the electronic conductivity and a fixed value of ionic conductivity due to Pb^{2+} ions and Ni^{2+} ions. As the nickel oxide content increases, more Ni^{2+} ions are released into the glass network. The hopping polarons, which participate in conduction, are grabbed by Ni^{2+} ions in pairs. This combination of cation–polaron pairs diffuse as single neutral units. Movement of these neutral units cannot result in electrical conductivity so the ac conductivity decreases as the content of NiO is increased from 0.2 to 1.0 mol%. The ionic conductivity of Pb^{2+} ions is constant in a high-temperature region. As a result of Austin and Mott's quantum mechanical tunneling model, the low-temperature part of conductivity can be explained [37]. From the results of dielectric properties, it is suggested that the significance of insulating nature is decreased with increasing content of nickel oxide.

4. Experimental Procedure

The following chemical formulas were used to synthesize the glasses for this study:

N_0: $40PbO-60GeO_2$
N_2: $40PbO-59.8GeO_2$: $0.2NiO$

N_4: 40PbO-59.6GeO_2: 0.4NiO
N_6: 40PbO-59.4GeO_2: 0.6NiO
N_8: 40PbO-59.2GeO_2: 0.8NiO
N_{10}: 40PbO-59.0GeO_2: 1.0NiO

High-purity chemicals of PbO, GeO_2, and NiO (all are in mol%) are taken in powder form in a suitable proportion and well grounded. In a controlled furnace, properly stirred mixtures were melted at 1000–1100 degrees Celsius until bubble-free liquid was obtained for about 1 h. As a result, the bubble-free melt was poured into rectangular brass molds and instantly annealed at 350 °C for four hours in another furnace for avoiding air cracks in the samples. At last, glass samples are obtained. The Philips expert system was used to obtain the XRD patterns of glass specimens. The programmable VIBRA HT kit was utilized to conduct density measurements, employing Archimedes' law and O-xylene as the buoyant liquid. The optically polished samples have dimensions of 1 cm × 1 cm × 0.1 cm. A high-precision NIR spectrophotometer was used to obtain optical spectra. FTIR spectra were obtained using a standard spectrophotometer by a powdered sample of KBr pellets. The Raman spectra, photoluminescence spectra, and DSC traces are produced on standard instruments. The glasses were silver-coated for measuring electrical parameters. The coating was then exposed to a blower until it was dried. The electrical parameters were taken on a prominent LCR meter. The precision in determining the ε' is $\frac{1}{10^3}$ and that of loss is $\frac{1}{10^4}$.

5. Conclusions

40PbO–60GeO_2: NiO glasses were prepared with different concentrations from 0.2 to 1.0 mol%. The XRD and DSC patterns indicated no crystalline phases. The IR and Raman spectral analysis revealed that the entropy of the glass network decreased with increasing content of nickel oxide. The investigation into the optical absorption, ESR, and photoluminescence peaks has demonstrated a growing occurrence of Ni^{2+} ions within tetrahedral sites. The values of susceptibility and magnetic moments of the glass samples have indicated Ni^{2+} ion transformation from octahedral to tetrahedral sites. The electrical parameters ε' and tan δ have increased and σ_{ac} increased whereas the activation energy has decreased, which proved their decreasing insulating nature with an increasing concentration of nickel oxide. Our observations are in agreement with prior studies on similar glass systems, where the introduction of transition metal oxides like NiO has been shown to enhance ionic conductivity due to the creation of non-bridging oxygen ions, which facilitate the movement of charge carriers. Therefore, an ionic conducting glass may be useful in solid electrolytes based on the studied glass system.

Author Contributions: L.V.: Conceptualization, writing—review and editing. S.M.S.: Writing, data curation. R.V.: Calculation and data procuring. K.P.: Formal analysis, editing. P.R.B.: Data analysis, editing. S.-J.K.: Formal analysis, conceptual discussion. G.N.R.: Supervision. All authors have read and agreed to the published version of the manuscript.

Funding: This work was supported by the 2024 Yeungnam University Research Grant.

Data Availability Statement: The data will be available on reasonable request from the corresponding author due to privacy.

Conflicts of Interest: The authors declare no conflicts of interest.

References

1. Rada, S.; Culea, M.; Neumann, M.; Culea, E. Structural role of europium ions in lead–borate glasses inferred from spectroscopic and DFT studies. *Chem. Phys. Lett.* **2008**, *460*, 196. [CrossRef]
2. Pisarski, W.; Goryczka, T.; Wodecka-Dus, B.; Plonska, M.; Pisarska, J. Structure and properties of rare earth-doped lead borate glasses. *Mater. Sci. Eng. B* **2005**, *122*, 94. [CrossRef]
3. El-Egili, K.; Doweidar, H.; Moustafa, Y.M.; Abbas, I. Structure and Some Physical Properties of PbO–P_2O_5 Glasses. *Phys. B* **2003**, *339*, 237. [CrossRef]

4. Marcel, P. Advanced glasses. *Ann. Chim. Sci. Matter* **2003**, *28*, 87.
5. Alderman, O.L.; Hannon, A.C.; Holland, D.; Umesaki, N. On the germanium–oxygen coordination number in lead germanate glasses. *J. Non-Cryst. Solids* **2014**, *386*, 56. [CrossRef]
6. Nouadji, M.; Ivanova, Z.G.; Poulain, M.; Zavadil, J.; Attaf, A. Glass formation, physicochemical characterization and photoluminescence properties of new Sb_2O_3–PbO–ZnO and Sb_2O_3–PbO–ZnS systems. *J. Alloy Compd.* **2013**, *549*, 158. [CrossRef]
7. Kashyap, R. Photosensitive Optical Fibers: Devices and Applications. *Opt. Fiber Technol.* **1994**, *1*, 17. [CrossRef]
8. Zmojda, J.; Kochanowicz, M.; Miluski, P.; Golonko, P.; Baranowska, A.; Ragi, T.; Dorosz, J.; Kuwik, M.; Pisarski, W.; Pisarska, J.; et al. Luminescent Studies on Germanate Glasses Doped with Europium Ions for Photonic Applications. *Materials* **2020**, *13*, 2817. [CrossRef] [PubMed]
9. Kuwik, M.; Kowalska, K.; Pisarska, J.; Pisarki, W.A. Spectroscopic properties of Pr^{3+}, Tm^{3+}, and Ho^{3+} in Germenate based Glass Systems modified by TiO_2. *Materials* **2023**, *16*, 61. [CrossRef]
10. Zelniok, D.; Cramer, C.; Eckert, H. Structure/Property Correlations in Ion-Conducting Mixed-Network Former Glasses: Solid-State NMR Studies of the System $Na_2O-B_2O_3-P_2O_5$. *Chem. Mater.* **2007**, *19*, 3162. [CrossRef]
11. Iezid, M.; Legouera, M.; Goumeidane, F.; Poulain, M.; Nazabal, V.; Lebullenger, R. Glass formation in the Sb_2O_3–$CdCl_2$–$SrCl_2$ ternary system. *J. Non-Cryst. Solids* **2011**, *357*, 2984. [CrossRef]
12. Loiko, P.; Dymshits, O.; Zhilin, A.; Alekseeva, I.; Yumashev, K. Influence of NiO on phase transformations and optical properties of ZnO–Al_2O_3–SiO_2 glass–ceramics nucleated by TiO_2 and ZrO_2. Part II. Optical absorption and luminescence. *J. Non. Cryst. Solids* **2013**, *376*, 99. [CrossRef]
13. Greaves, G.N.; Sen, S. Inorganic glasses, glass-forming liquids and amorphizing solids. *Adv. Phys.* **2007**, *56*, 1. [CrossRef]
14. Mysen, B.O.; Richet, P. *Silicate Glasses and Melts: Properties and Structure*; Elsevier: Amsterdam, The Netherlands, 2005.
15. Keppler, H.; Bagdassarov, N. The speciation of Ni and Co in silicate melts from optical absorption spectra to 1500 C. *Chem. Geol.* **1999**, *158*, 105. [CrossRef]
16. Reddy, M.S.; Krishna, G.M.; Veeraiah, N. Spectroscopic and magnetic studies of manganese ions in ZnO–Sb_2O_3–B_2O_3 glass system. *J. Phys. Chem. Solids* **2006**, *67*, 789. [CrossRef]
17. Nassau, K. The material dispersion zero in infrared optical waveguide materials. *Bell Syst. Technol. J.* **1981**, *60*, 327. [CrossRef]
18. Mc Cubbin, T.K., Jr.; Atmosoekarto, S.; Withstandley, V. Infrared Spectroscopy with a Copper-Doped Germanium Detector. *Appl. Opt.* **1967**, *6*, 1131. [CrossRef]
19. SrinivasaReddy, M.; Prasad, S.V.G.V.A.; Veeraiah, N. Valence and coordination of chromium ions in ZnO–Sb_2O_3–B_2O_3 glass system by means of spectroscopic and dielectric relaxation studies. *Phys. Status Solidi A* **2007**, *204*, 816. [CrossRef]
20. Tagiara, N.S.; Chatzipanagis, K.I.; Bradtmuller, H.; Rodrigues, A.C.M.; Moncke, D.; Kamitos, E.I. Network former mixing effects in alkali germanotellurite glasses: A vibrational spectroscopic study. *J. Alloys Compd.* **2021**, *882*, 160782. [CrossRef]
21. Rada, S.; Erhan, R.V.; Bodnarchuk, V.; Barbu-Tudoran, L.; Culea, E. SANS, RAMAN and SEM studies of lead-germanate glasses doped with the manganese oxide. *J. Alloys Compd.* **2021**, *882*, 160721. [CrossRef]
22. Mattos, G.R.S.; Bordon, C.D.S.; Kassab, L.R.P.; Issa, S.A.; ALMisned, G.; Tekin, H.O. Towards obtaining the optimum physical, optical, and nuclear radiation attenuation behaviours of tellurite-germanate glasses through Eu_2O_3 reinforcement: Glass synthesis, experimental and theoretical characterization study. *Ceram. Int.* **2023**, *49*, 986. [CrossRef]
23. Pipes, R.S.; Shelby, J.E. Formation and properties of soda lime germanate glasses. *J. Non-Cryst. Solids* **2021**, *553*, 120506. [CrossRef]
24. Falci, R.F.; Guerineau, T.; Delarosbil, J.L.; Messaddeq, Y. Spectroscopic properties of gallium-rich germane gallate glasses doped with Tm^{3+}. *J. Lumin.* **2022**, *249*, 119014. [CrossRef]
25. Jianga, X.P.; Yang, Z.M.; Liu, S.H. Energy transfer between Yb^{3+} and Er^{3+} in barium gallogermanate glass. *J. Appl. Phys.* **2009**, *105*, 103113. [CrossRef]
26. Pisarska, J.; Sołtys, M.; Górny, A.; Kochanowicz, M.; Zmojda, J.; Dorosz, J.; Dorosz, D.; Sitarz, M.; Pisarski, W.A. Rare earthdoped barium gallo-germanate glasses and their near-infrared luminescence properties. *Spectrochim. Acta A* **2018**, *201*, 362–366. [CrossRef] [PubMed]
27. Saffarini, G.; Saiter, J.M.; Matthiesen, J. Thaermal stability and percolation threshold of Ge-Se-Fe glasses. *Mater. Lett.* **2007**, *61*, 432–436. [CrossRef]
28. Gandhi, Y.; Mohan, N.K.; Veeraiah, N. Role of nickel ion coordination on spectroscopic and dielectric properties of ZnF_2–As_2O_3–TeO_2: NiO glass system. *J. Non-Cryst. Solids* **2011**, *357*, 1193. [CrossRef]
29. Kumar, G.R.; Rao, M.K.; Srikumar, T.; Rao, M.C.; Kumar, V.R.; Veeraiah, N.; Rao, C.S. Spectroscopic, dielectric dispersion and dc conductivity studies of Sb_2O_3 doped lithium fluoroborophosphate glasses mixed with small concentrations of NiO. *J. Alloys Compd.* **2018**, *752*, 179. [CrossRef]
30. Teja, P.V.; Rajyasree, C.; Krishna, S.M.; Tirupataiah, C.; Rao, D.K. Effect of some VA group modifiers on R_2O_3 (R = Sb, Bi)–ZnF_2–GeO_2 glasses doped with CuO by means of spectroscopic and dielectric investigations. *Mater. Chem. Phys.* **2012**, *133*, 239. [CrossRef]
31. Beattie, I.R.; Livingston, K.M.S.; Ozin, G.A.; Renolds, D.J. Single-crystal Raman spectra of arsenolite (As_4O_6) and senarmonite (Sb_4O_6). The gas-phase Raman spectra of P_4O_6, P_4O_{10}, and As_4O_6. *J. Chem. Soc. A* **1970**, *449–451*. [CrossRef]
32. Galeener, F.L.; Geissberger, A.E.; Ogar, G.W.; Loehman, R.E. Vibrational dynamics in isotopically substituted vitreous GeO_2. *Phys. Rev. B* **1983**, *28*, 4768. [CrossRef]
33. Lee, J.D. *Concise Inorganic Chemistry*, 5th ed.; Blackwell Science Ltd.: Hoboken, NJ, USA, 1996.

34. Brown, A.; Gupta, V. Coordination environments of Ni^{2+} in germinate glass matrices. *Opt. Mater.* **2018**, *36*, 1262–1266.
35. Suresh, B.; Reddy, M.S.; Reddy, A.S.S.; Gandhi, Y.; Kumar, V.R.; Veeraiah, N. Spectroscopic features of Ni^{2+} ion in PbO–Bi_2O_3–SiO_2 glass system. *Spectrochim. Acta Part A Mol. Biomol. Spectrosc.* **2015**, *141*, 263. [CrossRef] [PubMed]
36. Gao, G.; Reibstein, S.; Spiecker, E.; Peng, M.; Wondraczek, L. Broadband NIR photoluminescence from Ni^{2+}-doped nanocrystalline Ba–Al titanate glass ceramics. *J. Mater. Chem.* **2012**, *22*, 2582. [CrossRef]
37. Austin, I.G.; Mott, N.F. Polarons in crystalline and non-crystalline materials. *Adv. Phys.* **1969**, *18*, 657. [CrossRef]

inorganics

MDPI

Article

Discovering Novel Glass with Robust Crystallization Resistance via Amorphous Phase Separation Engineering

Mou Deng [1], Mingzhong Wang [1,2], Yu Rao [1], Yinsheng Xu [3,*], Dong Wu [1], Shisheng Lin [4] and Ping Lu [1,*]

1 School of Materials Science and Engineering, Wuhan University of Technology, Wuhan 430070, China; 268778@whut.edu.cn (M.D.); wangmz@csholding.com (M.W.); 245800@whut.edu.cn (Y.R.); dongwu@whut.edu.cn (D.W.)
2 Yichang CSG Photovoltaic Glass Co., Ltd., Yichang 443000, China
3 State Key Laboratory of Silicate Materials for Architectures, Wuhan University of Technology, Wuhan 430070, China
4 College of Physics and Energy, Fujian Normal University, Fuzhou 350117, China; linshisheng@fjnu.edu.cn
* Correspondence: xuyinsheng@whut.edu.cn (Y.X.); lupingwh@whut.edu.cn (P.L.)

Abstract: Amorphous phase separation (APS) is ubiquitously found in a large number of glass systems, because the glass can be regarded as solid with a heterogeneous structure at the nanoscale. However, little attention has been paid to the big challenges in utilizing APS in searching novel amorphous glass from above to below, which highlights the meticulous microstructure tunability of glass. Correspondingly, we develop a novel SiO_2-Al_2O_3-P_2O_5-Li_2O-ZrO_2 glass with APS (SAPLZ APS) which has robust crystallization resistance via the APS engineering. A comparative study is conducted to reveal the APS–crystallization property relationship. It can be found that the introduced APS can substantially impede the precipitated crystal growth in the studied glass system. Considering detailed glassy structure and microstructure, a diffusion barrier around each Li-rich droplet is created by the presence of P^{5+} concentration surrounding the Li-rich region. Meanwhile, due to the increase in Q^4 at the expense of Q^3, the polymerization degree in the Si-rich amorphous area can be enhanced, further increasing its viscosity and raising the kinetic barrier of Si-related crystal growth. These findings provide a new manner to develop new glass with superior anti-crystallization performance.

Keywords: glass; amorphous phase separation; glass ceramics; crystallization performance

1. Introduction

Glass ceramics, as an attractive and versatile material, are generated via controlled crystallization to precipitate microcrystals/nanocrystals within a glass matrix [1,2]. The properties of GCs are largely determined by the properties and composition of their crystalline phase. Lithium aluminum silicate (LAS) GCs have been widely studied because of their low coefficient of thermal expansion and high mechanical strength and transparency [3–6]. In this regard, the process of glass crystallization involves two distinct stages: nucleation and crystal growth. Amorphous phase separation (APS) plays a significant role in nucleation and has following influences on the subsequent crystal growth.

From the past until now, and even in the future, as the inherent character of glass (perceived as a "super-cooled liquid") changes, studies on APS have been and will be crucial in the field of glass science [7–12]. In fact, since glass can be regarded as solid with a heterogeneous structure at the nanoscale, APS is ubiquitously found in a large number of glass systems (e.g., fluoroaluminosilicate glass, fluorosilicate glass, halidephosphate glass, and chalcogenide glass). In terms of silicate glass, its APS should be caused by the occurrence of competition between different cations for oxygen ions; whereas the bridging oxygen ions are arranged around silicon ions in the form of silicon–oxygen tetrahedra, the non-bridging oxygen ions are pulled by network modifiers or intermediate ions to satisfy their respective structural arrangements [13]. When P_2O_5 is added to the glass, it tends to phase separate the glass and form enriched domains with distinct glassy components by removing non-bridging, oxygen-forming lithium ions from the silicate network [14].

Prof. Zanotto measured the nucleation kinetics of the barium disilicate crystal phase in BaO-SiO_2 glass utilizing a quantitative optical microscope in the temperature range of 673–807 °C in 1986 [15]. It was found that the nucleation rate of amorphous phase separation glass is much lower than that of 33.1% mol% BaO glass, which is close to the composition without phase separation. However, the effect of phase separation on crystallization nucleation is substantial but indirect, due to changes in the composition of the barium-rich phase that undergoes crystallization nucleation. Therefore, after 1 h of heat treatment at low temperatures, the separation of amorphous phases can be ignored. The higher the BaO content in the glass, the higher the nucleation rate of the crystals, which is closer to the stoichiometric composition of disilicates. After heat treatment at higher temperatures where phase separation occurs, the nucleation rate of crystals in phase-separated glass tends to converge due to the very similar matrix composition in the glass at a given temperature. Prof. Zanotto released a significant paper in 2020 [16], which reviewed the most applicable results of a research project on the synchronous crystallization kinetics analysis of APS, BaO-SiO_2, and Li_2O-SiO_2 glass, and supplemented recent research findings. The authors consider the following: First, the influence of the droplet interface can be neglected, as they are two orders of magnitude smaller compared to the crystal/nucleus surface energy. Second, the crystal nucleation kinetics of the phase-separating glasses exhibit an initial increase and subsequently reach a constant value as the composition of the liquid matrix attains the binodal boundary. Finally, the APS behavior drives the composition of the glass matrix towards the stoichiometric composition of the barium disilicate crystal phase, thereby enhancing the crystal nucleation rate, but it never reaches the value of the stoichiometric composition of the barium disilicate crystal glass. These comprehensive results clearly indicate that the primary function of APS is to transform the composition of the glass matrix into a stoichiometric crystal phase composition, thereby leading to enhanced crystal nucleation. These discoveries have solved a long-standing mystery in glass science.

Prof. Zanotto demonstrated the ubiquity of this phenomenon, showing the possible effect of APS. There are generally two significant impacts of APS: promoting nucleation and forming self-organization [17]. The former shifts the composition of the separating glass towards that of the stoichiometric crystal phase, where APS acts as a special catalyst for nucleation [18]. The latter forms the diffusion barrier around each APS droplet, inhibiting crystal growth tendency [19]. As such, when precursor glass is subjected to phase separation treatment, a discernible reduction in the size of crystalline grains will be observed, which is beneficial to controlling crystallization and obtaining nanometer-sized crystals in glass [20]. Great endeavors have been undertaken to reveal APS [21–23]. Despite their successes, little attention has been paid to the big challenges in utilizing APS engineering in searching novel amorphous glass from above to below, which highlights the meticulous microstructure tunability of glass.

Herein, we develop a new amorphous glass with robust crystallization resistance, the SiO_2-Al_2O_3-P_2O_5-Li_2O-ZrO_2 glass with APS (abbreviated as SAPLZ APS), via in situ APS of the glass matrix. A comprehensive study was conducted to unearth the APS type, glassy structure, microstructure, and crystallization performance in the studied glass system. The controlled APS progress leads to the highly ordered, Li-rich glassy particles (surrounded by P^{5+}) and Si-rich glassy components in the glass network, forming a diffusion barrier around each Li-rich droplet and facilitating an enhanced polymerization degree in the Si-rich amorphous area, thus achieving the ingenious control of glassy structure.

2. Experiment

2.1. Synthesis

The precursor glass (PG) with the nominal composition (in mol.%) of 75.34SiO_2-8.00Al_2O_3-1.00P_2O_5-15.67Li_2O-2.0ZrO_2 (abbreviated as SAPLZ glass) was prepared via the conventional melt-quenching route. A trace of P_2O_5 was used as a nucleation agent to induce phase separation. The preparation process of the precursor glass was as follows. First, we calculated the amount of each oxide required based on the chemical formula of the target crystal. Secondly, we considered which raw materials to utilize to introduce the required oxides. Finally, we calculated the required mass of raw material. This experiment utilized high-quality analytical grade reagents such as SiO_2 (≥99.98%), Al_2O_3 (≥99.8%), $NH_4H_2PO_4$ (≥99.0%), Li_2CO_3 (≥99.9%), and ZrO_2 (≥99.8%) as raw materials. Weigh the reagents with an electronic balance, placing a new weighing paper on the balance to protect it, and perform a tare operation before each weighing. When weighing, it is advisable to add reagents in small amounts and multiple times. Excess samples should not be placed back into the original kit. The operation should be conducted in a stable environment. The errors of the sample should be controlled below 2 mg. Then, transfer the sample to an agate mortar for mixing and grinding for 15 min. During this process, it is necessary to ensure that the sample is initially mixed evenly and the particle size of the reagents is controlled within the same range. Afterwards, use a dry ball mill to stir the raw materials for 1 h. When removing the sample, it is necessary to use dust-free paper to collect as many residual reagents as possible from the machine. After mixing evenly, gradually transfer the mixture to a Pt-Rh (90/10 wt%) crucible. During the transfer process, there are two aspects to be aware of. First, do not shake or vibrate the crucible to avoid the raw materials becoming too compact in the crucible. If the raw materials are too compact when transferred to a high-temperature furnace, it may cause the raw materials to erupt and potentially change the composition. Second, the raw materials should be added to the crucible up to two-thirds of its volume, and if there are too many volatile components in the raw materials, it is best not to exceed one-half of the volume. The melting process of the sample was as follows: we melted the first batch of raw materials added to the crucible at 1500 °C for 10 min, and then removed the crucible and added the remaining raw materials. We repeated the above operation until all the raw materials were added. After preliminary melting at 1500 °C, we heated up to 1620 °C and melted for 1 h under an atmospheric environment. The molten glass was then quickly poured onto a copper plate and rapidly quenched. Another copper plate was used to quickly press the glass. If the glass naturally fractured, the fragments were directly collected and transferred to the crucible. If the glass remained intact, it was crushed, and the glass fragments and powder were completely collected and transferred to the crucible. This step helped to ensure that the glass liquid discharged bubbles and melted evenly. The glass was then placed in a high-temperature furnace and melted at 1620 °C for 2 h under an atmospheric environment. At the end of the operation, the glass melt was poured into a copper mold and pressed between two copper plates. Subsequently, the obtained bulk glasses were cut into square coupons and polished for the following heat treatment. In the first step, the glasses were subjected to phase separation at 600 °C/620 °C/640 °C for different periods of time to produce the APS. In the second step, the samples were further heat-treated at 750 °C, 775 °C, 800 °C for 1, 2, or 4 h to research their crystallization ability. Meanwhile, the as-quenched glass specimens were also heat-treated at 700 °C, 725 °C, 750 °C, 775 °C, 800 °C, and 825 °C for 1 h as reference. The heat treatment procedure for the samples was as follows: the muffle furnace was inspected to ensure it was clean and free of any other samples. The samples were placed in a ceramic boat and positioned at the center of the muffle furnace to ensure uniform heating. The heating program was set with a ramp rate of 10 °C/min. After holding at the target temperature for a sufficient duration, the samples were allowed to cool down to room temperature within the furnace. All heat treatments in this experiment were conducted using the same equipment to avoid temperature variations from different devices. Prior to the experiments, the muffle furnace temperature was set to 600 °C and 700 °C, and the actual temperatures were measured using a thermocouple. The results showed that the actual temperatures were close to the set values, allowing us to rule out any interference from the equipment with the experimental results.

2.2. Characterization

Differential scanning calorimetry (DSC) data of the samples were collected using a Simultaneous Thermal Analyzer (STA 449 F1, NETZSCH, Selb, Germany) in the temperature range of 40–900 °C with a heating rate of 10 °C/min under a constant flow of nitrogen gas. The glass transition temperature T_g and crystallization temperature T_p were measured from the DSC curves. The sample preparation for DSC was as follows: first, the sample was cleaned with ultrasonic deionized water for 5 min, then dried for 20 min, and finally ground into a powder with a particle size of 10–80 microns using an agate mortar. The crucible used in the DSC curve test was a platinum crucible. The phases identified in the samples were observed by X-ray diffraction (XRD, PANalytical X'Pert Pro, Almelo, The Netherlands) using a Cu Kα radiation source at 40 kV and 40 mA and scanning from 5° to 80° at the scanning speed of 10°/min. The samples for XRD testing were prepared using a standardized method. First, the samples were crushed in a mold, then transferred to an agate mortar and ground for ten minutes to ensure that the samples were dry and had a particle size of 10–80 μm. Field emission SEM (Hitachi S-4800, 5 kV, Ibaraki, Japan) was used to observe the crystal morphology and size after crystallization. The SEM sample preparation process was as follows: The sample was placed in mold and crushed. Small glass pieces with smooth and flat surfaces and appropriate dimensions were selected. The fracture surface of the glass was etched with 2 vol% aqueous hydrofluoric acid (HF) for 30 s, cleaned by ultrasonication deionized water for 5 min, and then dried for 20 min (as water on the surface of the sample would seriously affect the test results and the pictures taken would have many stripes). Finally, the glass sputter coated with around 4 nm thick platinum to avoid the charging effect during the SEM analysis. The heat-treated glass samples were characterized using a JEM-1400Plus field emission transmission electron microscope (TEM). Additionally, the corresponding selected area diffraction (SAED) was performed to determine whether crystals had precipitated. The specimens for TEM and SAED analysis were prepared by bombarding the sample with argon gas as accelerating ions to sputter the surface atoms using a multifunctional beam ion dilution device (Gatan PIPS 695, Pleasanton, CA, USA). The thickness of the specimen is 30–40 nm. ^{29}Si and ^{31}P MAS NMR spectra were obtained using 4 mm probes on Bruker AVANCE spectrometers operating at a frequency of 500 MHz. The sample preparation for NMR was the same as that for XRD, and it was possible to directly use the samples prepared for XRD testing. Both XRD and NMR are non-destructive tests, so they would not affect the experimental results. ^{29}Si and ^{31}P MAS NMR spectra were obtained using 4 mm probes on Bruker AVANCE spectrometers operating at a frequency of 500 MHz. The ^{29}Si MAS NMR experiments utilized a rotation rate of 6.0 kHz with a resonance frequency of 99.4 MHz. The pulse length for the 90° excitation was set to 6 μs, and a cyclic delay of 300 s was utilized. As for ^{31}P MAS NMR, a rotation rate of 12.0 kHz and a resonance frequency of 202.5 MHz were applied. The 90° pulse length was set to 2.8 μs for quenched samples and 2.9 μs for heat-treated samples. The cycle delay was adjusted to 80 s for quenched samples and 160 s for heat-treated samples.

3. Results and Discussion

The samples were fabricated via a melt-quenching–thermal annealing route which is easily scalable for mass industrial production. XRD analyses on samples with various heat treatment times at 620 °C all show the amorphous hump coming from the glassy structure (Figure 1a). Transmission electron microscopy (TEM) discloses that the spherical particles of size of 5–9 nm are homogenously distributed among the glass matrix (Figure 1b). The corresponding selected area electron diffraction (SAED) shows a halo ring (a characteristic of an amorphous phase structure), verifying the occurrence of APS. The ^{31}P magic angle spinning nuclear magnetic resonance (MAS-NMR) spectroscopy reveals huge transformation of the local environment surrounding P before/after being heat-treated at 620 °C and illustrates a broad resonance peak which can be well deconvoluted into three sub-bands (originated from $Q^1{}_2$, $Q^1{}_1$, and Q^0 species, respectively) [24,25], as presented in Figures 1c and 2 and Table 1.

$Q^n{}_m$ is the structural unit of [PO_4] (n represents the number of P atoms connected to each [PO_4]; m represents the number of Al atoms attached to each [PO_4], and $m + n \leq 4$). With the annealing time prolonged from 0.5 h to 4 h, Q^0 and $Q^1{}_1$ species with lots of non-bridging oxygens gradually dominate, which gives evidence of the formation of relative isolated [PO_4]-composed groups surrounding Li-rich liquid–liquid immiscibility [26]. Differential scanning calorimeter (DSC) analysis on PG indicates a relatively minor exothermic peak at 645 °C, followed by three distinct and prominent exothermic peaks observed at temperatures of 704 °C, 750 °C, and 825 °C, respectively, while the exothermic peak at 645 °C should be assigned to the APS, since it nearly disappears upon preheating at 620 °C (Figure 1d). Based on more careful observations, with the periods of heat treatment time for APS prolonged, the exothermic peaks (at 704 °C, 750 °C, and 825 °C) slightly shift toward a low temperature. Meanwhile, due to the progressive changes in APS, the glass transition temperature (T_g) increases, which indicates an elevated viscosity and improved glass-forming ability [27]. Combined with the amorphous hump from their X-ray diffraction (XRD) patterns (Figure 1a), a conclusion can be drawn that the APS effect indeed exerts influences on the crystallization kinetics in SAPLZ glass.

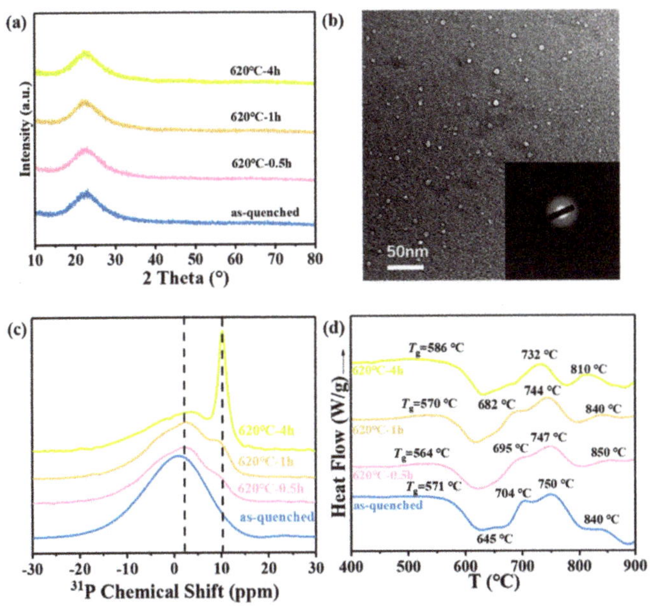

Figure 1. (**a**) XRD patterns of SAPLZ APS glass. (**b**) TEM image of SAPLZ APS glass; insets are the corresponding FFT pattern. (**c**) Solid-state, single-pulse ^{31}P MAS-NMR spectra of SAPLZ glass before/after APS. (**d**) DSC curve of the SAPLZ glass before/after APS recorded at a heating rate of 10 K/min (T_g denotes the glass transition temperature).

To provide more insights into the local structural transformation in glass before/after APS, ^{29}Si MAS-NMR spectra are exhibited in Figure 3 and Table 2. Obviously, samples exhibit the amorphous nature. Each spectrum can be fitted into three Gaussian peaks, representing the contribution of Q^2, Q^3, and Q^4 species (denoted as Q^n, where n represents the number of bridging oxygen atoms for each [SiO_4]) [28,29]. As expected, Q^2 remains relatively stable, Q^3 declines, and Q^4 increases, which indicates the formation of Si-rich glassy components as well as the enhancement of polymerization degree within the glass network, and coincides with the results from the DSC curves (Figure 1d). Based on the discussions above, the valuable insight about the connectivity in the glass structure can offer crucial information with respect to the type of APS (containing an Li-rich phase surrounded by P^{5+} and a Si-rich phase).

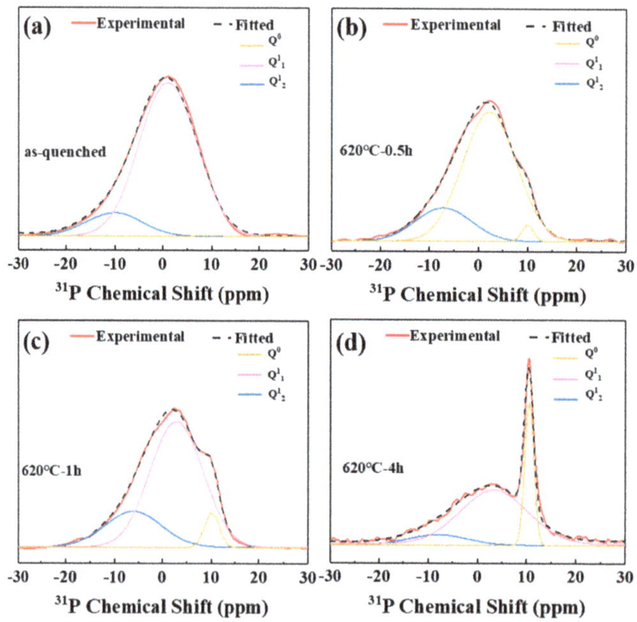

Figure 2. Gaussian fitting of the ^{31}P MAS-NMR spectra of (**a**) as-quenched PG, (**b**) 620 °C-0.5 h, (**c**) 620 °C-1 h, and (**d**) 620 °C-4 h.

Table 1. The ^{31}P chemical shifts δiso (±0.5 ppm) and relative area (±2%) extracted from MAS-NMR spectra.

Sample	As-Quenched PG	620 °C-0.5 h	620 °C-1 h	620 °C-4 h
Q^0 δ_{iso}/ppm	9.14	9.93	10.01	10.42
$Q^1{}_1$ δ_{iso}/ppm	0.86	2.13	2.62	3.26
$Q^1{}_2$ δ_{iso}/ppm	−10.22	−7.25	−6.6	−8.12
Q^0 Relative area/%	0.34	2.23	5.53	23.59
$Q^1{}_1$ Relative area/%	86.70	76.67	73.81	65.29
$Q^1{}_2$ Relative area/%	12.96	21.09	20.66	11.13

The SEM observation on SAPLZ glass annealed at 620 °C and 750 °C for different times show their microstructural evolution (Figure 4). It can be seen that the number and size of the precipitated crystal particles show perceivable increases with the annealing duration time at 750 °C prolonged (Figure 4a–c). Fortunately, the crystallinity deviates substantially toward reduction with the APS-producing time increasing from 1 h to 4 h, confirming its inhibitory effect on crystal growth (Figure 4d–i). Specifically, there is a substantial reduction in the number of in situ precipitate crystals in 1 h pre-APS, and only a minimal amount of crystals can be attained after preheating at 620 °C for 4 h. Crystal growth is substantially impeded when the SAPLZ glass undergoes APS, whereby longer APS time leads to a stronger inhibitory effect on crystallization, which provides the evidence for the discovery of a novel amorphous glass with wonderful anti-crystallization performance via the APS process.

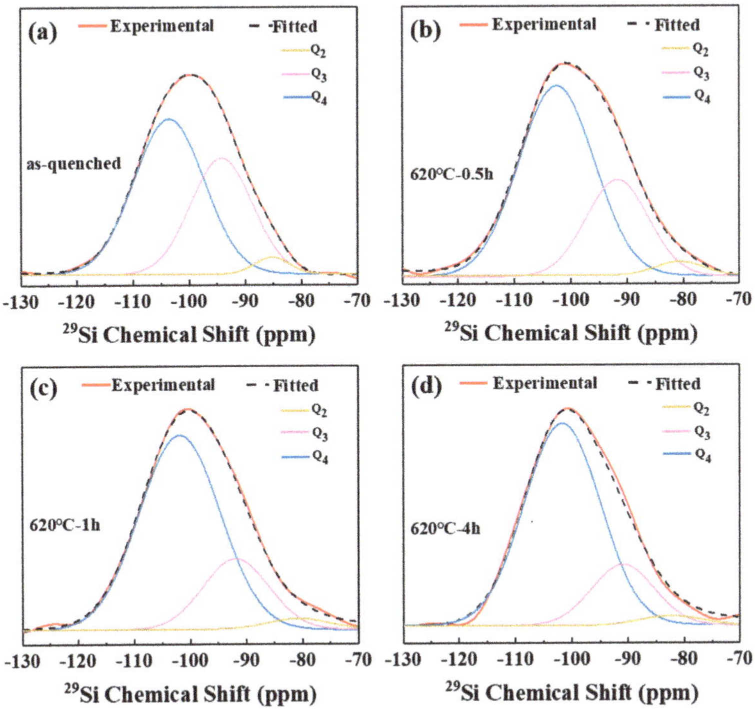

Figure 3. Gaussian fitting of the ^{29}Si MAS-NMR spectra of (**a**) as-quenched PG, (**b**) 620 °C-0.5 h, (**c**) 620 °C-1 h, and (**d**) 620 °C-4 h.

Table 2. The ^{29}Si chemical shifts δiso (±0.5 ppm) and relative area (±2%) extracted from MAS NMR spectra.

Sample	As-Quenched PG	620 °C-0.5 h	620 °C-1 h	620 °C-4 h
$Q^{(2)}$ δ_{iso}/ppm	−85.38	−80.90	−80.85	−82.37
$Q^{(3)}$ δ_{iso}/ppm	−94.30	−91.86	−91.98	−91.00
$Q^{(4)}$ δ_{iso}/ppm	−103.62	−102.60	−101.91	−101.74
$Q^{(2)}$ Relative area/%	3.04	3.01	2.96	3.17
$Q^{(3)}$ Relative area/%	38.71	28.95	22.28	19.45
$Q^{(4)}$ Relative area/%	58.25	68.04	74.76	77.38

To search the novel amorphous glass with strong crystallization resistance, the detailed effect of APS on SAPLZ glass crystallization performances are surveyed. When the heat treatment temperature increases to 725 °C, the phase transformation from amorphous to crystalline (tetragonal SiO_2) occurs (Figure 5a), at which point the crystallinity is 2.14%. When the heat treatment temperature further increases from 775 °C to 825 °C, the signal of tetragonal $LiAlSi_2O_6$ becomes gradually stronger for the samples, and the crystallinity of the sample reaches 74.63%. Fortunately, after introducing the APS process, the adverse crystallization process is effectively restrained (Figure 5b), for instance, at a crystallization temperature of 750 °C and a phase separation temperature of 600 °C, the crystallinity was 1.32%. At a phase separation temperature of 640 °C, the crystallinity increased to 10.07%. However, at a phase separation temperature of 620 °C, no crystal nucleation was observed. The final APS temperature is optimized to 620 °C, based on the consideration of its best crystallization resistance behavior. As illustrated in Figure 5c, the SAPLZ APS glass exhibits superior anti-crystallization performance: the heat treatment time is ~4 h at

750 °C to a reach crystallinity of ~45% (in comparison, the heat-treatment time is only ~1 h for SAPLZ glass). When we thermally anneal the samples at 620 °C for different sintering times and then heat-treat them at 750 °C for 1 h, the X-ray diffraction peaks intensity is reduced with the increased APS time, and correspondingly, the crystallinity decreases from 43.7% to 0 via calculating the crystallization fraction (Figure 5d and Table 3), which demonstrates the well-suppressed crystal growth effect from APS for SAPLZ glass. Based on the results above, it can be seen that a kind of novel amorphous material possessing robust crystallization resistance, SAPLZ APS glass (620 °C-4 h), is successfully fabricated via APS engineering.

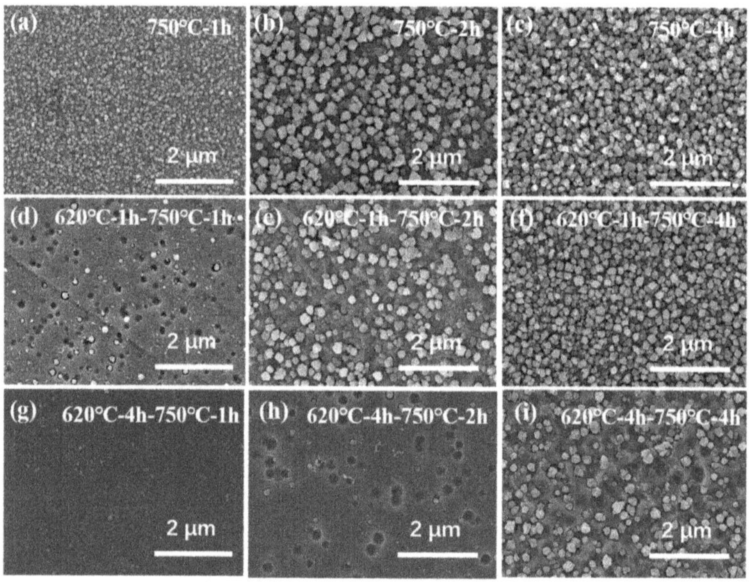

Figure 4. SEM observation on the (**a**) 750 °C-1 h, (**b**) 750 °C-2 h, (**c**) 750 °C-4 h, (**d**) 620 °C-1 h-750 °C-1 h, (**e**) 620 °C-1 h-750 °C-2 h, (**f**) 620 °C-1 h-750 °C-4 h, (**g**) 620 °C-4 h-750 °C-1 h, (**h**) 620 °C-4 h-750 °C-2 h and (**i**) 620 °C-4 h-750 °C-4 h.

Table 3. The crystallization fraction of different SAPLZ glass after heat treatment at 620 °C for 0 h/0.5 h /1 h/2 h/4 h and 750 °C for 1 h.

Sample	Crystallinity (%)
620 °C-0 h and 750 °C-1 h	43.7
620 °C-0.5 h and 750 °C-1 h	22.1
620 °C-1 h and 750 °C-1 h	7.2
620 °C-2 h and 750 °C-1 h	1.3
620 °C-4 h and 750 °C-1 h	0

The possible mechanism for the inhibitory effect of APS on precipitated crystal in the studied glass system is proposed, as illustrated in Figure 6. To be noted, the presence of P^{5+} concentration surrounding the Li-rich region can be attributed to propensity of P^{5+} for dissociation from the glassy network structure, which leads to the formation of isolated $[PO_4]$ with an affinity for Li^+ [12,19]. Subsequently, Li+ can effectively capture P^{5+} during APS, thus forming a diffusion barrier around each Li-rich droplet to inhibit further Li-related crystal growth. Meanwhile, the APS induces an increase in Q^4 at the expense of Q^3, which implies an enhanced polymerization degree in the Si-related amorphous area and subsequently increases its viscosity and raises the kinetic barrier of Si-related crystal growth.

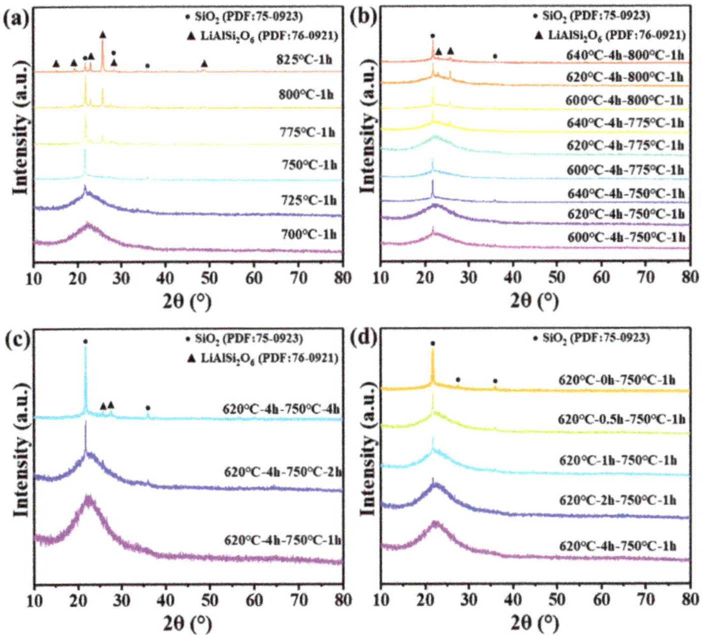

Figure 5. XRD patterns of SAPLZ glass heat-treated (**a**) from 700 °C to 825 °C for 1 h, (**b**) at 600 °C/620 °C/640 °C for 4 h and 750 °C/775 °C/800 °C for 1 h, (**c**) at 620 °C for 4 h and 750 °C for 1 h/2 h/4 h, and (**d**) at 620 °C for 0 h/0.5 h/1 h/2 h/4 h and 750 °C for 1 h.

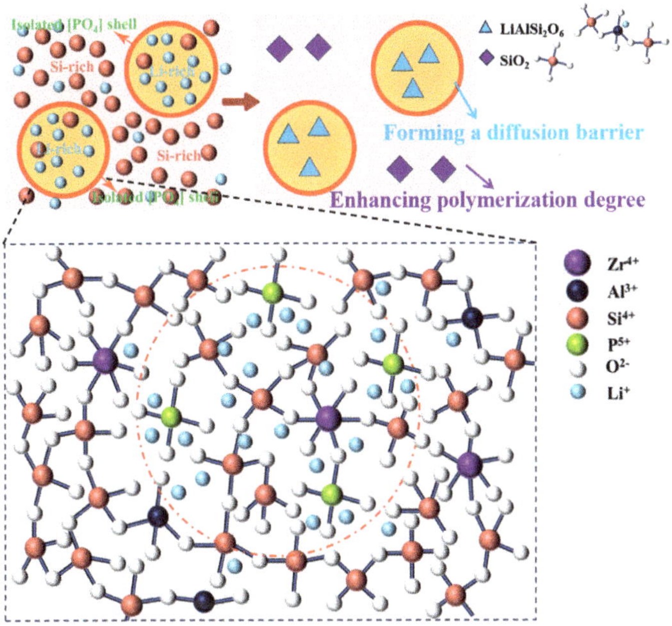

Figure 6. Schematic illustration of possible mechanism for the inhibitory effect of APS on precipitated crystal in the studied glass system.

4. Conclusions

In summary, we have proposed a new SAPLZ APS glass with robust crystallization resistance for the first time, which highlights the meticulous microstructure tunability of glass. The structure analysis shows that the APS is attributed to the Li-rich phase (surrounded by P^{5+}) and Si-rich phase. Probing into the crystallization performance of SAPLZ APS glass, anti-crystallization performance greatly improved upon introducing APS engineering, which reveals the inhibitory effect of APS on crystal growth in the studied glass system. Not only was a diffusion barrier around each Li-rich droplet formed, but the polymerization degree was increased as well in the Si-related amorphous area. This study demonstrates an effective application of APS engineering in developing new glass.

Author Contributions: M.D. contributed to conceptualization, experimental design and performance, writing, and funding acquisition; M.W. contributed to resources and project administration; Y.R. contributed to methodology and text editing; Y.X. contributed to funding acquisition, conceptualization, and supervision; D.W. contributed to investigation and data curation; S.L. contributed to funding acquisition, conceptualization, and supervision; P.L. contributed to supervision and conceptualization. All authors have read and agreed to the published version of the manuscript.

Funding: This research was supported by the National Natural Science Foundation of China (No.: U2241236, 52372014, 12304442), the Key R&D Project of Hubei Province (2022BAA025) and Natural Science Foundation of Fujian Province (2022J05091).

Data Availability Statement: Data are contained within the article.

Conflicts of Interest: Mingzhong Wang was employed by the Yichang CSG Photovoltaic Glass Co., Ltd. The remaining authors declare that the research was conducted in the absence of any commercial or financial relationships that could be construed as a potential conflict of interest. The Yichang CSG Photovoltaic Glass Co., Ltd. company was not involved in the study design, collection, analysis, interpretation of data, the writing of this article or the decision to submit it for publication.

References

1. Rawlings, R.; Wu, J.; Boccaccini, A. Glass-ceramics: Their production from wastes—A review. *J. Mater. Sci.* **2006**, *41*, 733–761. [CrossRef]
2. Zhao, M.; Cao, J.; Wang, Z.; Li, G. Insight into the dual effect of Fe_2O_3 addition on the crystallization of CaO-MgO-Al_2O_3-SiO_2 glass-ceramics. *J. Non-Cryst. Solids* **2019**, *513*, 144–151. [CrossRef]
3. Zheng, W.; Lin, M.; Cheng, J. Effect of phase separation on the crystallization and properties of lithium aluminosilicate glass-ceramics. *Glass Phys. Chem.* **2013**, *39*, 142–149. [CrossRef]
4. Sakamoto, A.; Himei, Y.; Hashibe, Y. β-Spodumene glass-ceramic with anomalous low thermal expansion. *Adv. Mater. Res.* **2008**, *39*, 381–386. [CrossRef]
5. Laczka, M.; Laczka, K.; Cholewa-Kowalska, K.; Kounga, A.B.; Appert, C. Mechanical properties of a lithium disilicate strengthened lithium aluminosilicate glass-ceramic. *J. Am. Ceram. Soc.* **2014**, *97*, 361–364. [CrossRef]
6. Xia, L.; Yang, Y.; Zhang, X.; Zhang, J.; Zhong, B.; Zhang, T.; Wang, H.; Wen, G. Crystal structure and wave-transparent properties of lithium aluminum silicate glass-ceramics. *Ceram. Int.* **2018**, *44*, 14896–14900. [CrossRef]
7. Rafferty, A.; Hill, R.; Wood, D. Amorphous phase separation of ionomer glasses. *J. Mater. Sci.* **2000**, *35*, 3863–3869. [CrossRef]
8. Rafferty, A.; Hill, R.; Wood, D. An investigation into the amorphous phase separation characteristics of an ionomer glass series and a sodium-boro-silicate glass system. *J. Mater. Sci.* **2003**, *38*, 2311–2319. [CrossRef]
9. Chen, Y.; Liu, S.; Zhou, Y.; Shang, P.; Shan, Z.; Zhang, J. Effect of Al_2O_3 content on amorphous phase-separation and self-limited crystallization of phosphosilcate glasses. *J. Non-Cryst. Solids* **2022**, *584*, 121505. [CrossRef]
10. Zandonà, A.; Moustrous, M.; Genevois, C.; Véron, E.; Canizarès, A.; Allix, M. Glass-forming ability and ZrO_2 saturation limits in the magnesium aluminosilicate system. *Ceram. Int.* **2022**, *48*, 8433–8439. [CrossRef]
11. Lin, C.; Bocker, C.; Rüssel, C. Nanocrystallization in oxyfluoride glasses controlled by amorphous phase separation. *Nano Lett.* **2015**, *15*, 6764–6769. [CrossRef] [PubMed]
12. Liu, S.J.; Zhang, Y.F.; He, W.; Yue, Y.Z. Transparent phosphosilicate glasses containing crystals formed during cooling of melts. *J. Non-Cryst. Solids* **2011**, *357*, 3897–3900. [CrossRef]
13. Kirkpatrick, R.J.; Brow, R.K. Nuclear magnetic resonance investigation of the structures of phosphate and phosphate-containing glasses: A review. *Solid State Nucl. Magn. Reson.* **1995**, *5*, 9–21. [CrossRef] [PubMed]
14. Li, J.; Wang, M.; Lu, P. Nucleating role of P_2O_5 in nepheline-based transparent glass-ceramics. *J. Am. Ceram. Soc.* **2021**, *104*, 5614–5624. [CrossRef]

15. Zanotto, E.D.; James, P.F.; Craievich, A.F. The effects of amorphous phase separation on crystal nucleation kinetics in BaO-SiO_2 glasses. *J. Mater. Sci.* **1986**, *21*, 3050–3064. [CrossRef]
16. Zanotto, E.D. Effect of liquid phase separation on crystal nucleation in glass-formers. Case closed. *Ceram. Int.* **2020**, *46*, 24779–24791. [CrossRef]
17. CAHN, J.W. The Metastable Liquidus and Its Effect on the Crystallization of Glass. *J. Am. Ceram. Soc.* **1969**, *52*, 118–121. [CrossRef]
18. James, P.F. Glass ceramics: New compositions and uses. *J. Non-Cryst. Solids* **1995**, *181*, 1–15. [CrossRef]
19. Zhu, L.; Wang, M.; Xu, Y.; Zhang, X.; Lu, P. Dual effect of ZrO_2 on phase separation and crystallization in Li_2O-Al_2O_3-SiO_2-P_2O_5 glasses. *J. Am. Ceram. Soc.* **2022**, *105*, 5698–5710. [CrossRef]
20. Nicoleau, E.; Schuller, S.; Angeli, F.; Charpentier, T.; Jollivet, P.; Le Gac, A.; Fournier, M.; Mesbah, A.; Vasconcelos, F. Phase separation and crystallization effects on the structure and durability of molybdenum borosilicate glass. *J. Non-Cryst. Solids* **2015**, *427*, 120–133. [CrossRef]
21. James, P.F. Liquid-phase separation in glass-forming systems. *J. Mater. Sci.* **1975**, *10*, 1802–1825. [CrossRef]
22. Sycheva, G.A. Phase separation and crystallization in glasses of the lithium silicate system $xLi_2O \cdot (100 - x)SiO_2$ (x = 23.4, 26.0, 33.5). *Glass Phys. Chem.* **2011**, *37*, 135–149. [CrossRef]
23. Bussey, J.M.; Weber, M.H.; Smith-Gray, N.J.; Sly, J.J.; McCloy, J.S. Examining phase separation and crystallization in glasses with X-ray nano-computed tomography. *J. Non-Cryst. Solids* **2023**, *600*, 121987. [CrossRef]
24. Zhang, L.; Eckert, H. Short-and medium-range order in sodium aluminophosphate glasses: New insights from high-resolution dipolar solid-state NMR spectroscopy. *J. Phys. Chem. B* **2006**, *110*, 8946–8958. [CrossRef] [PubMed]
25. Lang, D.P.; Alam, T.M.; Bencoe, D.N. Solid-state $^{31}P/^{27}Al$ and $^{31}P/^{23}Na$ TRAPDOR NMR investigations of the phosphorus environments in sodium aluminophosphate glasses. *Chem. Mater.* **2001**, *13*, 420–428. [CrossRef]
26. Yu, X.; Wang, M.; Rao, Y.; Xu, Y.; Xia, M.; Zhang, X.; Lu, P. Unveiling the evolution of early phase separation induced by P_2O_5 for controlling crystallization in lithium disilicate glass system. *J. Eur. Ceram. Soc.* **2023**, *43*, 5381–5389. [CrossRef]
27. Micoulaut, M. The slope equations: A universal relationship between local structure and glass transition temperature. *Eur. Phys. J. B—Condens. Matter Complex Syst.* **1998**, *1*, 277–294. [CrossRef]
28. Bradtmüller, H.; Gaddam, A.; Eckert, H.; Zanotto, E.D. Structural rearrangements during sub-T_g relaxation and nucleation in lithium disilicate glass revealed by a solid-state NMR and MD strategy. *Acta Mater.* **2022**, *240*, 118318. [CrossRef]
29. Iqbal, Y.; Lee, W.; Holland, D.; James, P. Metastable phase formation in the early stage crystallisation of lithium disilicate glass. *J. Non-Cryst. Solids* **1998**, *224*, 1–16. [CrossRef]

MDPI AG
Grosspeteranlage 5
4052 Basel
Switzerland
Tel.: +41 61 683 77 34

Inorganics Editorial Office
E-mail: inorganics@mdpi.com
www.mdpi.com/journal/inorganics

www.ingramcontent.com/pod-product-compliance
Lightning Source LLC
Chambersburg PA
CBHW062034220926
51850CB00010B/2543